Fracture Mechanics and Fatigue Design in Metallic Materials

Fracture Mechanics and Fatigue Design in Metallic Materials

Editor

Dariusz Rozumek

MDPI • Basel • Beijing • Wuhan • Barcelona • Belgrade • Manchester • Tokyo • Cluj • Tianjin

Editor
Dariusz Rozumek
Opole University of Technology
Poland

Editorial Office
MDPI
St. Alban-Anlage 66
4052 Basel, Switzerland

This is a reprint of articles from the Special Issue published online in the open access journal *Metals* (ISSN 2075-4701) (available at: https://www.mdpi.com/journal/metals/special_issues/fracture_mechanics_fatigue_design_metals).

For citation purposes, cite each article independently as indicated on the article page online and as indicated below:

LastName, A.A.; LastName, B.B.; LastName, C.C. Article Title. *Journal Name* **Year**, *Volume Number*, Page Range.

ISBN 978-3-0365-2730-7 (Hbk)
ISBN 978-3-0365-2731-4 (PDF)

© 2021 by the authors. Articles in this book are Open Access and distributed under the Creative Commons Attribution (CC BY) license, which allows users to download, copy and build upon published articles, as long as the author and publisher are properly credited, which ensures maximum dissemination and a wider impact of our publications.

The book as a whole is distributed by MDPI under the terms and conditions of the Creative Commons license CC BY-NC-ND.

Contents

About the Editor vii

Preface to "Fracture Mechanics and Fatigue Design in Metallic Materials" ix

Ashutosh Sharma, Min Chul Oh and Byungmin Ahn
Recent Advances in Very High Cycle Fatigue Behavior of Metals and Alloys—A Review
Reprinted from: *Metals* **2020**, *10*, 1200, doi:10.3390/met10091200 1

Wenjie Wang, Jie Yang, Haofeng Chen and Qianyu Yang
Capturing and Micromechanical Analysis of the Crack-Branching Behavior in Welded Joints
Reprinted from: *Metals* **2020**, *10*, 1308, doi:10.3390/met10101308 25

Wei Xu, Yanguang Zhao, Xin Chen, Bin Zhong, Huichen Yu, Yuhuai He and Chunhu Tao
An Ultra-High Frequency Vibration-Based Fatigue Test and Its Comparative Study of a Titanium Alloy in the VHCF Regime
Reprinted from: *Metals* **2020**, *10*, 1415, doi:10.3390/met10111415 39

Garikoitz Artola and Javier Aldazabal
Hydrogen Assisted Fracture of 30MnB5 High Strength Steel: A Case Study
Reprinted from: *Metals* **2020**, *10*, 1613, doi:10.3390/met10121613 55

Behnam Zakavi, Andrei Kotousov and Ricardo Branco
The Evaluation of Front Shapes of Through-the-Thickness Fatigue Cracks
Reprinted from: *Metals* **2021**, *11*, 403, doi:10.3390/met11030403 71

Grzegorz Lesiuk, Hryhoriy Nykyforchyn, Olha Zvirko, Rafał Mech, Bartosz Babiarczuk, Szymon Duda, Joao Maria De Arrabida Farelo and Jose A.F.O. Correia
Analysis of the Deceleration Methods of Fatigue Crack Growth Rates under Mode I Loading Type in Pearlitic Rail Steel
Reprinted from: *Metals* **2021**, *11*, 584, doi:10.3390/met11040584 85

Mansur Ahmed, Md. Saiful Islam, Shuo Yin, Richard Coull and Dariusz Rozumek
Fatigue Crack Growth Behaviour and Role of Roughness-Induced Crack Closure in CP Ti: Stress Amplitude Dependence
Reprinted from: *Metals* **2021**, *11*, 1656, doi:10.3390/met11101656 99

Seulbi Lee, Hanjong Kim, Seonghun Park and Yoon Suk Choi
Fatigue Variability of Alloy 625 Thin-Tube Brazed Specimens
Reprinted from: *Metals* **2021**, *11*, 1162, doi:10.3390/met11081162 111

Cameron R. Rusnak and Craig C. Menzemer
Fatigue Behavior of Nonreinforced Hand-Holes in Aluminum Light Poles
Reprinted from: *Metals* **2021**, *11*, 1222, doi:10.3390/met11081222 123

Kenichi Ishihara, Hayato Kitagawa, Yoichi Takagishi and Toshiyuki Meshii
Application of an Artificial Neural Network to Develop Fracture Toughness Predictor of Ferritic Steels Based on Tensile Test Results
Reprinted from: *Metals* **2021**, *11*, 1740, doi:10.3390/met11111740 137

Chi Liu, Liyong Ma, Ziyong Zhang, Zhuo Fu and Lijuan Liu
Research on the Corrosion Fatigue Property of 2524-T3 Aluminum Alloy
Reprinted from: *Metals* **2021**, *11*, 1754, doi:10.3390/met11111754 149

Dariusz Rozumek
Fracture Mechanics and Fatigue Design in Metallic Materials
Reprinted from: *Metals* **2021**, *11*, 1957, doi:10.3390/met11121957 . . . . **165**

About the Editor

Dariusz Rozumek, since 2021, has been a full professor at the Department of Mechanics and Machine Design of the Faculty of Mechanical Engineering from the Opole University of Technology, following 26 years of research and teaching activity at the Opole University of Technology, Poland. Since 1995, he has been a teacher and since 1998 he has been a researcher at the Opole University of Technology. In 2004 and 2005, he was guest researcher at the Department of Management and Engineering of the University of Padova (Vicenza, Italy). He obtained a PhD degree in Mechanics and Machine Design from the Opole University of Technology in 2002 and a DsC degree in Mechanics from the Czestochowa University of Technology in 2010. His research interests lie in the area of multiaxial fatigue and fracture including stress and energy approaches, welding and many other topics related to the field of mechanical engineering. He has published more than 275 scientific publications, eleven patents, eight book editions and thirteen book chapters. He participated in more than 10 European and National Projects. He is also a Guest Editor of several issues in various Journals and Reviewer of articles in 25 Journals. He was a co-organizer of several International Conferences of Fracture Mechanics and Material Fatigue as well as Sessions and Minisymposia. He is Professor of Undergraduate and Postgraduate programs with scientific fields in Advanced Strength of Materials, Mechanics, Theory of Mechanisms and Machines. Prof. Rozumek is a member of Section of Experimental Methods of Committee on Mechanics of the Polish Academy of Sciences, Polish Society of Theoretical and Applied Mechanics, Polish Group of Fracture Mechanics, European Structural Integrity Society (ESIS), a member of the working group in TC03 ESIS.

Preface to "Fracture Mechanics and Fatigue Design in Metallic Materials"

This Special Issue explores the most recent research advancements in the field of various design elements, with particular emphasis on their safety and reliability. The analysis included the study of initiation and fatigue crack growth in metallic materials.

This Special Issue has attracted submissions from Australia, Bangladesh, China, Ireland, Japan, Korea, Poland, Portugal, Spain, Ukraine, UK, and USA: 19 submissions were received and 12 articles were published.

Sharma et al. presented the review article entitled "Recent advances in very high cycle fatigue behavior of metals and alloys—A review", in which they review the research and development in the field of fatigue damage, focusing on the very high cycle fatigue (VHCF) of metals, alloys and steels. In addition, they showed the influence of various defects, crack initiation sites, fatigue models and simulation studies to understand the crack development in VHCF regimes.

Wang et al. presented the paper "Capturing and micromechanical analysis of the crack-branching behavior in welded joints", where the branching criterion or the mechanism governing the bifurcation of a crack in welded joints is analyzed. In their work, three kinds of crack-branching models that reflect simplified welded joints were designed.

Xu's group presented research entitled "An ultra-high frequency vibration-based fatigue test and its comparative study of a titanium alloy in the VHCF regime", which proposes an ultra-high-frequency (UHF) fatigue test of a titanium alloy TA11 based on an electrodynamic shaker to develop a feasible testing method in the VHCF regime.

Artola et al. conducted a scientific study entitled "Hydrogen assisted fracture of 30MnB5 high strength steel: A case study", where they investigated the impact of quench and tempering and hot-dip galvanizing on the hydrogen embrittlement behavior of high-strength steel. Slow-strain-rate tensile testing was employed to assess this influence.

Zakavi et al. presented a study entitled "The evaluation of front shapes of through-the-thickness fatigue cracks", in which new tools were shown to evaluate the crack front shape of through-the-thickness cracks propagating in plates under quasi-steady-state conditions.

Lesiuk, together with a group of co-authors, presents an article entitled "Analysis of the deceleration methods of fatigue crack growth rates under mode I loading type in pearlitic rail steel", in which they presented a comparison of the results of the fatigue crack growth rate for raw rail steel, steel reinforced with composite material—CFRP—and the case of counteracting crack growth using the stop-hole technique, as well as with an "anti-crack growth fluid".

Ahmed et al. in the presented work "Fatigue crack growth behaviour and role of roughness-induced crack closure in CP Ti: Stress amplitude dependence", where they investigated the fatigue crack propagation mechanism of CP Ti at various stress amplitudes and crack closure.

Another study entitled "Fatigue variability of alloy 625 thin-tube brazed specimens" was introduced by Lee et al., where they conducted fatigue tests at room temperature and 1000 K for alloy 625 tubes. The variability in fatigue life was investigated by analyzing the locations of the fatigue failure, fracture surfaces and microstructures of the brazed joint and tube.

Rusnak et al. presented the paper "Fatigue behavior of nonreinforced hand-holes in aluminum light poles", in which they fatigue tested nine poles with 18 openings using four-point bending at various stress ranges. Finite element analysis was used to complement the experimental study.

Ishihara et al. presented a study entitled "Application of an artificial neural network to develop fracture toughness predictor of ferritic steels based on tensile test results", where they analyzed the structural integrity of ferritic steel structures subjected to large temperature variations, which required the collection of the fracture toughness of ferritic steels in the ductile to brittle transition region.

Liu et al. presented the paper "Research on the corrosion fatigue property of 2524-T3 aluminum alloy", in which subjected the 2524-T3 aluminum alloy to fatigue tests under the conditions of R = 0, a 3.5% NaCl corrosion solution and loading cycles of 10^6, and the S-N curve was obtained. The fatigue source characterized by cleavage and fracture mainly comes from corrosion pits.

Finally, a summary of all the presented work is shown in the Editorial article.

Dariusz Rozumek
Editor

Review

Recent Advances in Very High Cycle Fatigue Behavior of Metals and Alloys—A Review

Ashutosh Sharma [1,2], Min Chul Oh [3] and Byungmin Ahn [1,2,]*

1. Department of Materials Science and Engineering, Ajou University, Suwon 16499, Korea; ashu@ajou.ac.kr
2. Department of Energy Systems Research, Ajou University, Suwon 16499, Korea
3. Metal Forming Technology R&D Group, Korea Institute of Industrial Technology, Incheon 21999, Korea; mc0715hj@kitech.re.kr
* Correspondence: byungmin@ajou.ac.kr; Tel.: +82-31-219-3531; Fax: +82-31-219-1613

Received: 14 August 2020; Accepted: 2 September 2020; Published: 8 September 2020

Abstract: We reviewed the research and developments in the field of fatigue failure, focusing on very-high cycle fatigue (VHCF) of metals, alloys, and steels. We also discussed ultrasonic fatigue testing, historical relevance, major testing principles, and equipment. The VHCF behavior of Al, Mg, Ni, Ti, and various types of steels were analyzed. Furthermore, we highlighted the major defects, crack initiation sites, fatigue models, and simulation studies to understand the crack development in VHCF regimes. Finally, we reviewed the details regarding various issues and challenges in the field of VHCF for engineering metals and identified future directions in this area.

Keywords: fatigue; fracture; very-high cycle; high-entropy alloy; powder metallurgy; fish eye

---

1. Introduction

1.1. History of Fatigue/Background

Preliminary observations were recorded at the beginning of the 19th century during the industrial revolution in Europe. During this time, several railways, heavy-duty locomotives, and engines accidentally failed after a long period of time. In 1829, W.A.S. Albert noticed this failure while performing cyclic loading on iron chain [1,2]. Later, in 1837, he reported a relation between cyclic load and lifespan of metal in a magazine. Following this observation, a cast-iron axle designer, J.V. Poncelet, used the term "fatigare" and F. Brainthwaite in Great Britain coined it as fatigue in 1854 [3,4].

In 1842, one of the worst rail disasters happened near Versailles, France. Several locomotives' axles broke on the way. After inspection by W.J.M. Rankine from British railways, a brittle fracture in the axle was confirmed [2]. Following this observation, some pioneering work performed by August Wöhler on the failure of locomotive axles built the foundation of fatigue understanding. Wöhler plotted the Krupp axle steel data with respect to stress (S) and number of cycles to failure (N). This plot was later named the S-N diagram [5,6]. The S-N diagram is useful for forecasting the fatigue life and endurance limit of metals, i.e., the limiting threshold value of stress below which an engineering material exhibits a high or infinite high fatigue life. Thus, A. Wöhler is regarded as the grandfather of modern fatigue technology [7].

In 1886, J. Bauschinger published the first investigation on the stress-strain behavior of materials under cyclic loading. At the end of the 19th century, Gerber and Goodman performed systematic parametric investigations and proposed simplified fatigue theories. In 1910, O.H. Basquin further proposed the shape of the S-N diagram by applying Wöhler's data on a log–log scale. Following various investigations, fracture mechanics was born via the crack propagation theory of A.A. Griffith in 1920 [8]. Various Manson-Coffin-Basquin (MCB) models and Langer models came into focus to study the strain characteristics which depend upon time. The MCB model was developed for tension-compression fatigue [9,10]. Later, the MCB model was replaced with Langer and Kandil and another model developed by Kurek and Lagoda under multiaxial loading [11–13]. Other fatigue investigations related to full-range fatigue regimes for more than 10^8 cycles (Kohout-Vechet model) were also pioneered including multi-fatigue damage parameters by recent researchers [14–16].

In practice, fatigue takes place under dynamic loading after a substantial period of service, in windmills, high-speed aircraft, ships, submarines, turbine blades, offshore platforms, launch vehicles, pressure vessels, etc. Most of the load-bearing applications in vehicles, engine parts, are loaded with 10^8 cycles, while railway components, bridges and wheels are loaded with 10^9 cycles in their lifetime. Therefore, there is a need for knowledge of fatigue behavior and safe operation limits in the VHCF regime. In the VHCF regime, the fatigue behavior is different compared to conventional high cycle fatigue. For instance, the fatigue behavior of high-strength steels in the VHCF regime show crack initiation at the inclusion site, while in high cycle fatigue (HCF), cracks are loacated preferentially at the surface [17,18]. Aluminium alloys have no fatigue limit but show slip band formation in VHCF regime [19,20]. There is no failure in Ti6Al4V at low stress ratios while the fatigue strength decreases at high stress ratios in VHCF [21–24]. It can be seen that VHCF behavior can be extrapolated from HCF data. The various distinct differences in VHCF and HCF have been reviewed by Li et al. [25]. VHCF mechanisms involving crack initiation at inclusion sites or other damage in the VHCF regime for defect free materials were reviewed by Zimmermann et al. [26]. In recent decades, the fatigue behavior of materials in VHCF regime has increased. However, the prominent cause of fatigue failure is still not clear and needs to be overviewed in detail. Therefore, this work focuses on the VHCF behavior of metals, alloys and steels. We also highlight the HCF fatigue behavior of novel high-entropy alloy systems for future guidance in extrapolating it to VHCF regime.

1.2. Classification

The word "fatigue" originated from the Latin word "fatigare" meaning "to tire". This means that materials fail sooner than expected after tiring under cyclic loading [3]. Fatigue is the most vital criteria for designing engineering materials. The severity of damage can be understood by the fact that fatigue occurs more often at lower stresses than at the yielding of materials [27]. Fatigue life depends on the number of cycles a material can withstand before fracture. The fracture can occur in a few cycles, or sometimes it takes a large number of cycles before fracture. However, before going further, we classify various fatigue processes occurring in engineering materials, as shown in Figure 1.

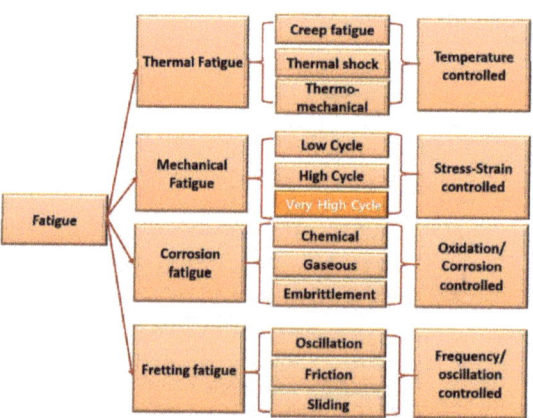

Figure 1. Classification of various types of fatigue failure occurring in engineering materials.

There are different types of fatigue failures according to the types of loading and conditions. Based on the fatigue life cycle, it can be divided according to the number of cycles needed to reach failure, such as (1) low cycle fatigue (LCF) (~10^4 ... ~10^5 cycles) and (2) high cycle fatigue (HCF) (~10^5 to 10^7 cycles). Additionally, fatigue life with more than 10^7 cycles is known as very-high cycle fatigue (VHCF) [28,29]. The fatigue crack initiation life is defined by the number of cycles required from the time of initiating a surface imperfection or grain size crack to the time of formation of a not-considerable length of crack. In other words, this is the number of cycles of fatigue crack propagation required for a microcrack to grow up to an easily definable length, 0.5 or 1 mm in length, while the number of cycles required to cause the failure of the material is the fatigue propagation life. The fatigue life is the sum of the fatigue crack initiation and propagation lives [30,31]. In this review, we mainly discuss VHCF.

Fatigue failure depends on other factors besides cyclic loads and temperature; these include the oxidizing environment, embrittlement, structural deformation, strain rate, and frequency of applied load. Various studies in the past have been devoted to LCF of engineering materials and the effect of temperature on fatigue life. The studies showed that fatigue life also depends on other factors such as strain rate, creep, oxidizing atmosphere, fretting cycle, frequency, and microstructural defects [30–35]. In fatigue failure, the dominating factor to damage is vague; hence, understanding of the failure mechanism is elusive.

1.3. Fatigue Testing Parameters

In fatigue testing, the load cycle is characterized by various parameters such as load ratio R ($\sigma_{min}/\sigma_{max}$), stress amplitude σ_a, mean stress σ_m, and stress range $\Delta\sigma$. Fatigue is evaluated in terms of fatigue strength and fatigue limit. Fatigue strength is the value of stress at which failure occurs after N_f cycles, while fatigue limit shows the limiting value of stress (S_f) at which failure occurs, as N_f becomes very large [27]. Fluctuating stresses can be classified into various categories. Commonly encountered stresses are tension-tension, compression-compression, pulsating, sinusoidal, irregular, or random stress cycles [8].

The fatigue data are generally represented by the stress (σ) versus the log number of cycles to failure (N). The S-N diagram is described for the value of the σ_m, or R (=−1), as shown in Figure 2. For Al alloys, there is no well-defined stress region below which these alloys fail. Therefore, the endurance limit is defined as the stress below which no failure occurs even for a large number of cycles (N_f), a measure of non-ferrous alloys [36].

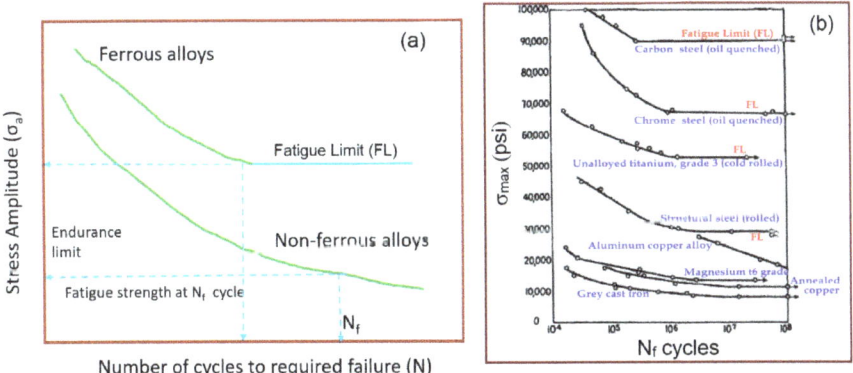

Figure 2. (a) Schematic of the S-N diagram and (b) S-N diagram of various metals and ferrous alloys [8].

The S-N diagram is employed to determine the number of cycles at a particular stress level before a fracture occurs [36]. It allows the engineers to design an alloy for the desired service life span, as shown in Figure 2a. Figure 2b presents the S-N diagram of various metals and alloys. We can see that mostly steels and titanium exhibit fatigue limit compared to ductile metals like Al or Cu. These diagrams only predict the fatigue limit of metals without any prior knowledge of cycles needed for crack nucleation and propagation, and the effect of sample dimensions. For ferrous alloys, there is a well-defined fatigue limit, but no such limit is defined for non-ferrous alloys. S-N diagrams, therefore, do not reveal the failure-free performance, hence more research is needed.

1.4. Fractography

In this section, we deal with a comparative morphology of LCF, HCF and VHCF fractured surfaces. The fatigue fractured surfaces after failure are assessed by various types of surface features termed as wavy beach patterns and striations, as shown in Figure 3 [37]. These features show the location of the crack tip that appears as concentric ridges away from the crack nucleation site. Beach patterns or clamshell marks can be seen with the naked eye and are of macroscopic dimensions. After analyzing the high-resolution images, we can see other secondary cracks and deep striation marks (Figure 3a–e).

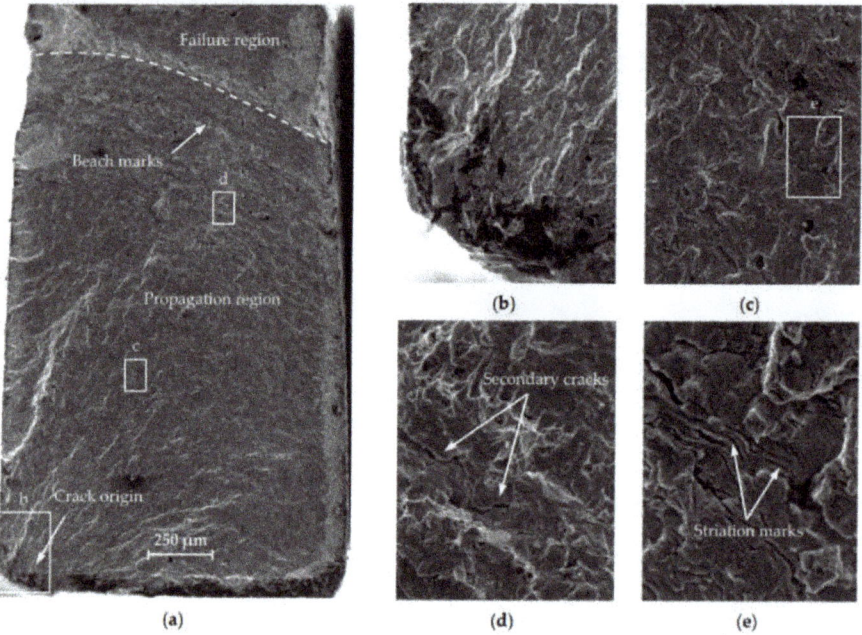

Figure 3. The fracture surface of S235JR steel shaft. (**a**) Overall surface; (**b**) origin of cracks; (**c**) propagation zone; (**d**) secondary cracks; (**e**) fatigue striations [37].

Figure 3 shows the various types of cracks induced during fatigue failure of S235JR steel shaft. Each beach pattern indicates a period during which the crack growth occurs, while the striation marks that appeared in the fatigue fractography are microscopic features and can only be seen with electron microscopy. Notably, thousands of striations exist within a beach pattern. The presence of these features confirms the fatigue failure, but their absence may or may not point to fatigue failure [8,37]. According to the Forsyth et al. [38], two different types of striations exist in the LCF regime, ductile light and dark bands, and brittle river like patterns. In the HCF regime of low-carbon steel, Kim et al. noticed a plastic flow induced by plastic deformation as compared to surface microcracks and microfissures in the LCF regime, as shown in Figure 4a–c [29].

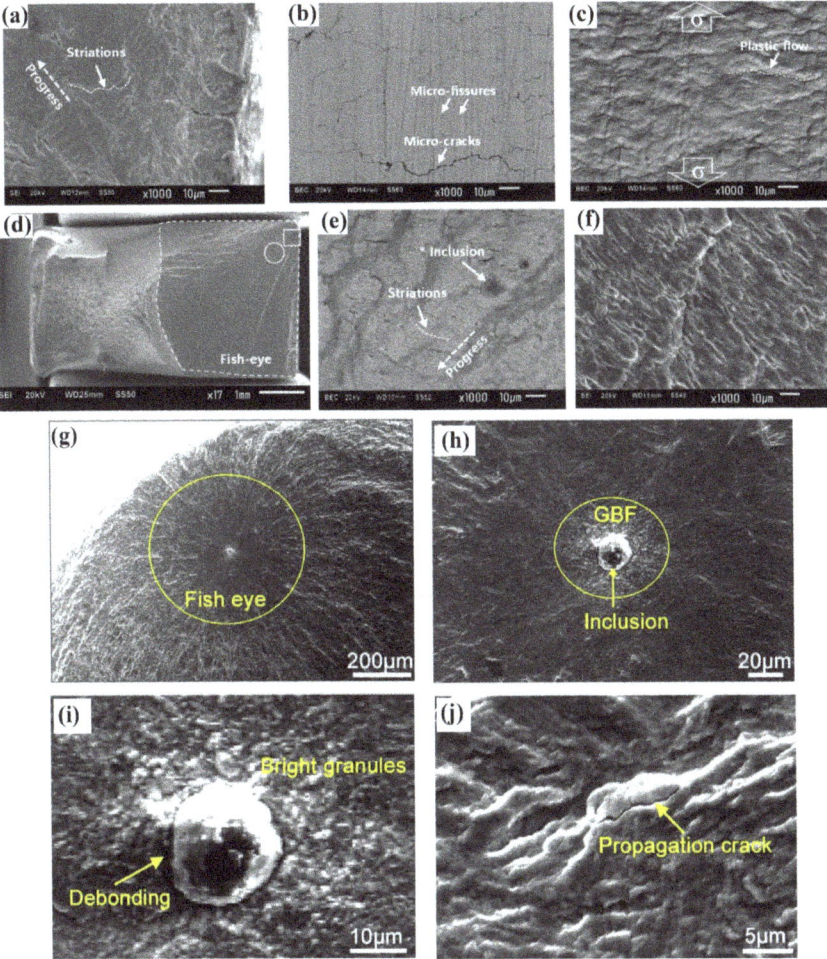

Figure 4. (**a**,**b**) The low cycle fatigue (LCF) fracture surface of low carbon steel tested at 2.0% strain, (**c**) HCF fracture surface of low carbon steel tested at 0.5% strain, (**d**) Overall surface including "white square" and "white circle" regions, (**e**) high resolution images of "white square" showing cracks and (**f**) "white circle" showing striations in high cycle fatigue (HCF) regime recorded from (**d**). The "white circle" in (**d**) shows fish-eye region observed at low resolution [29]. (**g**) A close view of fish-eye morphology, (**h**) granular bright facet (GBF) region and inclusion morphology, (**i**) high-resolution image of (**h**), and (**j**) fatigue crack propagation region. The fatigue failure was caused by inclusions in a bolt steel (σ_a = 600 MPa, N_f = 4.38 × 10^8) [39] (with permission from Elsevier, 2020).

Additionally, in the HCF regime, a "fish-eye" morphology was observed due to crack initiation at a CaO inclusions. The region under "white circle" corresponds to fish-eye (Figure 4d). Such fish-eye feature is frequently noticed in VHCF of heterogeneous materials containing foreign inclusions. Such inclusions originate during steel making from dephosphorization treatment. Distinct striations can be seen at the fish-eye surface (Figure 4f). A similar observation was observed by Zhao et al. during VHCF deformation of bolt steel [39]. The presence of inclusion was verified in their study at the site of the fish-eye region, which is 350–400 µm (Figure 4g). Further inspection revealed that the granular bright facet (GBF) region surrounding the fisheye was 70–100 µm (Figure 4h–i). The inclusion was mainly composed of Al_2O_3, MgO, CaO, etc., with a size of ~25 μm. The propagation region showed a wave like morphology (Figure 4j). Since we will focus on the VHCF behavior of materials, we will discuss VHCF testing and the behavior of common engineering metals and alloys systems in the following sections.

2. Very High Cycle Fatigue

With the continuous advancement in the high strength materials, the fatigue life of many instruments has exceeded 10^8 load cycles. VHCF is a major design issue in various fields of applications such as aircraft, auto parts, railways and jet engines. The fatigue life of various body parts like gas turbines discs (10^{10}), cylinder heads and blocks in a car engine (10^8), bearings, drilling equipment, engines of high-speed trains, and ships (10^9), falls under VHCF [40,41].

2.1. Conventional Fatigue Testing

Conventional fatigue testing systems operate at a frequency of 20 Hz and run for years to reach 10^{10} cycles while testing a specimen. Therefore, a design system for VHCF is required with a reliable S-N diagram. Piezoelectric transducers can be used to generate 20 kHz frequencies and can go up to 10^{10} in less than a week, but there is a disadvantage of increasing the temperature of the specimen during the test, hence cooling is required. In some steels, the fatigue difference between 10^7 and 10^{10} cycles can be over 200 MPa. Therefore, VHCF equipments are usually operated at lower stress to have a failure-free performance.

2.2. Ultrasonic Fatigue Testing

Ultrasonic fatigue test machines can provide the fatigue tests up to the VHCF range, which is the major scope of this review. The commonly investigated fatigue testing frequency for ultrasonic fatigue testing lies in the range of 15–30 kHz. Ultrasonic fatigue testing compresses the test time per cycle significantly, as shown in Table 1.

Table 1. Conventional and ultrasonic fatigue testing time for 10^{10} cycles [41].

Fatigue Test	Testing Time
Conventional fatigue tests (1 Hz)	320 years
Conventional fatigue tests (100 Hz)	3.2 years
Ultrasonic fatigue tests (20 kHz)	6 days

Hopkinson introduced the ultrasonic fatigue testing in the 20th century. He developed the electromagnetic system with a resonance frequency of 116 Hz, which can test fatigue at a maximum frequency of 33 Hz. Later in 1925, Jenkin tested the fatigue of Cu, Fe, and steel wires at 2.5 kHz. Following these observations, in 1929, Lehmann and Jenkin developed a pulsed air resonance system that can generate frequency up to 10 kHz. In 1950, Mason widened the scope of frequency window by the use of high-power ultrasonic waves for fatigue testing up to 20 kHz, which formed the first modern fatigue test equipment. Similarly, in the middle of the 20th century, higher frequencies were being employed for fatigue tests—92 kHz by Gerald in 1959, followed by 199 kHz by Kikukawa in 1965 [3].

Ultrasonic fatigue testing is a time-saving and cost-effective approach in the VHCF testing. The ultrasonic fatigue tester consists of a simple 20 kHz power generator and a piezoelectric converter which converts signals into ultrasonic waves of the same frequency [40]. Further, an ultrasonic horn amplifies these waves to get the desired stress amplitude, as shown in Figure 5a.

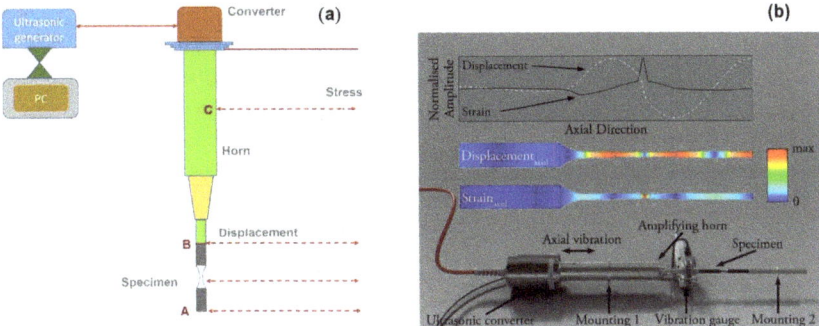

Figure 5. (a) Schematic of the ultrasonic fatigue test unit and stress displacement curve, and (b) Load train for R ≠ −1. Mechanical components and stress displacement curves are shown for =−1 when both mounting 1 and mounting 2 are omitted [42] (with permission from John Wiley and Sons, 2020).

The maximum stress is present in the central part of the specimen. The displacement becomes maximum at the specimen terminals (A and B). Other accessories include amplitude control units, oscilloscope, cycle counter, data acquisition systems like a camera, and displacement sensors. This form of ultrasonic fatigue testing has been used by various researchers under fully reversed tension (R = −1). With modern electromechanical or servo-hydraulic system attachments, such ultrasonic fatigue testing can be performed for gigacycle fatigue tests with a range of positive R ratios. A three-point bending test with R > 0 has been also used in the past [3]. Smooth or notched samples can be tested using these instruments under uniaxial stress. Torsion testing machines have also been designed that work like a uniaxial, under pulsed or continuous mode [43–45].

In addition, the ultrasonic fatigue test unit can also be used under various environments, such as air cooling, elevated temperature [46,47], cryogenic environment [46], or corrosive media [47,48]. When the loading cycles are superimposed (R ≠ −1), the sample is mounted on both sides such that the length of load train rods is equal to $\frac{1}{2}$ or one wavelength (Figure 5b). The mounting devices on both sides exert tensile or compressive stresses at nodes [48]. Such superimposed tests can be done by using the electromechanical or servo-hydraulic load frames to study the effect of different load ratios (R ≠ −1, −1, >0) on fatigue life [42,48].

3. Very High Cycle Fatigue of Engineering Materials

The VHCF behavior of different engineering materials varies. Therefore, the experimental database of various materials generated over the last decade is essential. Based on the literature data, the fatigue limit of some materials is in the range of 10^6–10^7 cycles. Besides, most of the high-strength materials exhibit a gradual loss of fatigue strength beyond 10^7 load cycle. The engineering materials in the VHCF regime can be broadly classified into two groups [40,41,49,50].

(1) Type 1: The gap of fatigue strength between 10^6 and 10^9 cycles is lower than 50 MPa; it includes ductile, homogeneous, and pure metals; e.g., Cu, Ni, and their alloys, low-carbon steels, spheroid cast iron, and some stainless steels;

(2) Type 2: In this category, the materials show a decrease in fatigue strength from 10^6 to 10^9 cycles. The decrease in strength can be up to 50–300 MPa; it includes most of the high strength steels, and most heterogeneous materials containing inclusions, pores, and a second phase that acts as a crack initiation site.

We will review these two classes of materials (Type 1 and Type 2) in the following sections, including the advanced high-entropy alloys (HEAs), briefly, and shed light on the S-N diagram as well as the fracture surfaces of different materials undergoing VHCF behavior.

3.1. Al Alloys

Al alloys are important for various manufacturing components in automobiles, space design, and naval architectures that are subjected to VHCF. Tschegg and Mayer investigated the fatigue behavior of AA 2024-T351 (Figure 6a). They used a testing frequency of 20 kHz under reversed load cycles in water [19]. They revealed that the fatigue fractures initiated at the surfaces of the specimens. The samples, tested in distilled water, caused a decrease in the fatigue strength of 30–40%. However, some of the specimens in water did not fracture at 72 MPa even beyond 10^9 cycles due to the presence of the alumina layer over the specimen surface. The cause of the failure was unclear, meaning it was not certain whether the alumina layer was beneficial for further corrosion protection or not.

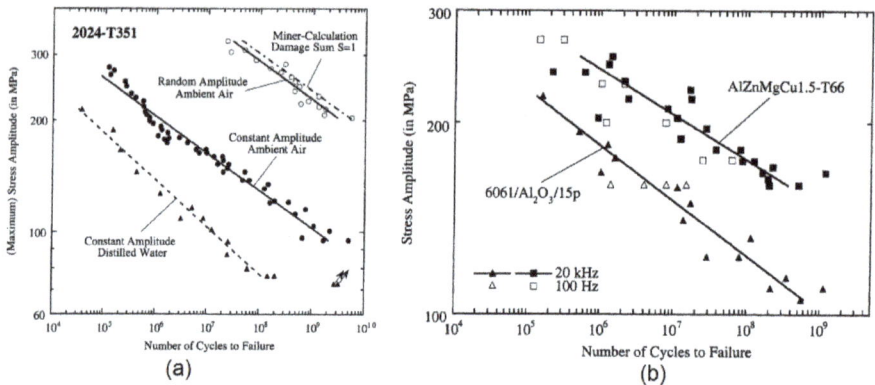

Figure 6. S-N diagram for (**a**) AA 2024-T351 (**b**) AlZnMgCu1.5-T66 and 6061/Al2O3/15p (the drawn lines show the 50% probability of failure at 20 kHz) [19] (with permission from Elsevier, 2020).

Further comparison of age-hardenable 7XXX and particle-reinforced 6XXX series Al composites showed no effect of cyclic frequency on lifetimes in VHCF regime, as shown in Figure 6b. Wang et al. studied the piezoelectric assisted VHCF behavior of 7075-T6, 6061-T6, and 2024-T3 alloys at 20 kHz [51]. Their observation showed that fracture took place right up to the 10^9 cycles at nominal cyclic stresses. The fracture surfaces showed voiding and faceting with discernible striation marks, except for AA2024-T3, which showed ductile tearing instead of striation.

Various researchers have investigated the heat treatment of alloys to improve fatigue strength. For example, Lee et al. investigated the VHCF behavior of a heat-treated non-Cu 7021 alloy at 20 kHz. The stress amplitude of peak-aged 7021 Al alloy was higher by 50 MPa than that of the solutionized alloy [52]. Oh et al. studied the HCF characteristics of AA 7075-T6 and 2024-T4 by shot peening. By fractographic analysis, they showed that 2024-T4 had the maximum fatigue life, while there was no improvement in fatigue life for 7075-T6. The reason was ascribed to the inside crack in 2024-T4 from shot peening while the crack was present on the surface for the 7075-T6 alloy. The un-peened specimens exhibited a similar trend [53].

Koutiri et al. studied the HCF of AlSi7Cu0.5Mg0.3 alloys. They reported that the presence of microstructural heterogeneities (IMCs, pores, Si particles, etc.) result in a different fatigue behavior. They also concluded that the biaxial tensile stress state is not detrimental to HCF [54].

3.2. Mg Alloys

Mg alloys have attracted attention due to their low weight, good machinability, and high specific strength and stiffness in many industries. Mg alloys find applications in structural parts of high-speed engines where fatigue life exceeds 10^8 cycles. Thus, the VHCF behavior of Mg and its alloys are essential for the high reliability of Mg components. Yang et al. reported the fatigue test of AZ31 alloy at 20 kHz ultrasonic frequency and R = −1 [55]. The fatigue strength was 89 MPa at 10^9 cycles without any fatigue limit (Figure 7a).

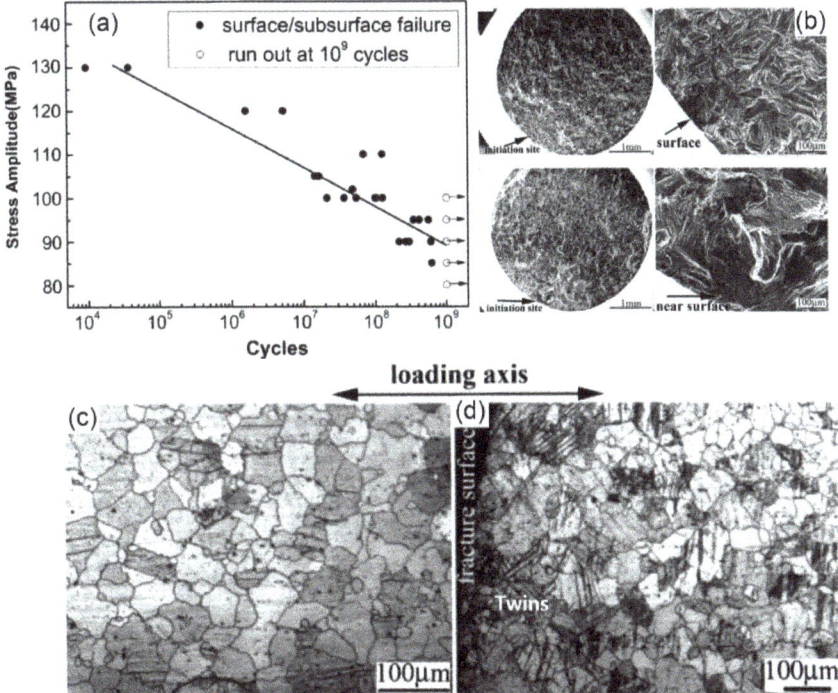

Figure 7. (**a**) S-N diagram of AZ31 Mg alloy tested at 20 kHz ultrasonic frequency and R = −1, (**b**) low and high-magnification scanning electron micrographs of the fracture surfaces of AZ31 alloy after ultrasonic fatigue, (**c**,**d**) the optical microstructure of the specimen section parallel to the gauge length before and after ultrasonic fatigue testing, respectively [55] (with permission from Elsevier, 2020).

Further analysis of the fracture surfaces (Figure 7b–d) showed that fatigue fracture originated beneath the surface, and the mechanism of failure was due to the twinning deformation, as shown in Figure 7d. Karr et al. employed a wrought AZ61 Mg alloy for ultrasonic fatigue testing in air. The fatigue strength at 10^9 cycles was 98 MPa. They reported that the failure mechanism was due to the formation of slip bands and their fracture with increasing load cycles at the surface [56].

The environmental effect on the fatigue behavior of Mg alloys was further studied in a study under low to high humidity (80% RH) and salt brine solution (5 wt.% NaCl). It was shown that the fatigue life of Mg alloys was reduced in high humidity and salt brine solutions due to the formation of corrosion pitting, while steel and Al alloy showed no influence of humidity on fatigue life. In comparison, there was a negligible effect on the fatigue life of steel and Al alloys. It was also shown that the application of chemical conversion coatings or anodizing can increase the fatigue life of Mg alloys [57].

Nascimetto et al. studied the influence of crystal texture on fatigue failure of extruded AZ31 and ZN11 Mg alloys. The AZ31 alloy was inhomogeneous and had strong fiber texture, which caused strong asymmetry in the tensile and compressive yield strengths. The metallographic observations revealed that the cracks are nucleated at the twin boundaries. Moreover, weakly textured and homogeneous ZN11 alloy showed no twinning, and the fatigue failure of ZN11 was initiated by cyclic slip deformation [58].

3.3. Cu Alloys

Pure Cu is a Type 1 material and has no common defects (pore, inclusion) for fatigue initiation in the VHCF range. We believe the fatigue fracture of Cu occurs at a threshold amplitude above which the formation of persistent slip bands occurs, leading to final shear and failure. These shear bands are not visible below the threshold near the VHCF range except for a minor surface roughening due to the irreversible cyclic slip component of dislocations [59–61].

Figure 8a shows the formation of surface projections in due course of embryonic development of crack for Type 1 materials. These rough surfaces later result in the formation of persistent slip bands that induce fatigue failure. Figure 8b shows the comparison of the S-N diagram for various types of Cu and ultrafine-grained Cu from the literature. In another study on microcrystalline for 10^9 cycles at stress amplitude (σ = 54 MPa), less stress was required to initiate slip band formation [60]. There was no instance of failure of the sample, even at the end of the 10^9 cycles. However, it was noticed that the visible surface projections grew to slip bands after 10^9 cycles. Similarly, Kunz et al. and Mughrabi et al. studied the VHCF behavior of ultrafine-grained Cu (300 nm) [60,61]. They found that the fatigue strength of ultrafine-grained Cu was twice that of large-grained Cu from 10^4 to 10^9 cycles. Contrariwise, high-purity ultrafine-grained Cu (99.99%) showed a lower fatigue strength, but it was still higher than that of large-grained Cu, as it is unstable at high cycles [60,61]. Figure 8b shows the comparison of the S-N diagram for commercial Cu and ultrafine-grained Cu [62,63].

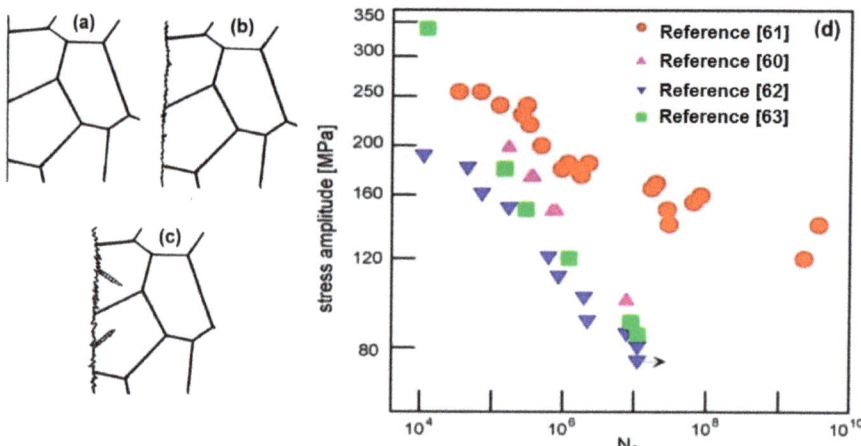

Figure 8. Schematic for gradual surface roughening during fatigue failure. (**a–c**) Initial, mid, and final surface roughening, respectively [59] (with permission from Elsevier, 2020); (**d**) a comparison of S-N diagram for various commercial ultrafine-grained Cu [61] (with permission from Elsevier, 2020).

3.4. Ni Alloys

Nickel and its alloys are primarily known for the high-temperature and oxidation-resistant applications in turbine blades and rotors which operate in the VHCF range. Bathias examined the VHCF behavior of a Ni-Cr-Co alloy (UDIMET 500). He found that there is no visible effect of frequency on fatigue in VHCF regime, though the fatigue strength showed a decreasing trend, as shown in Figure 9 [64].

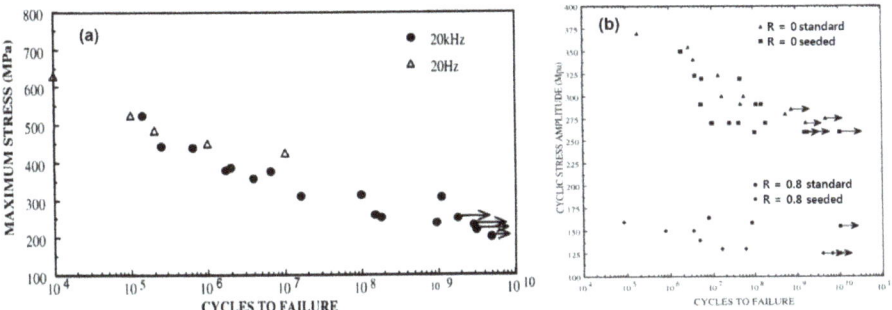

Figure 9. (a) S-N diagram for Udimet 500 alloy. R = −1, (b) N18 nickel alloy at 450 °C [64] (with permission from John Wiley and Sons, 2020).

In another example, high-temperature VHCF of N18 turbine disc alloys with inclusions were tested in the VHCF range at 450 °C. The inclusion containing seeds had lesser fatigue strength compared to unseeded ones, showing the destructive effect of inclusions in these alloys [64].

Chen et al. tested Inconel 718 for fatigue life at ambient temperature and R = −1. The S-N diagram is shown in Figure 10. They showed that below a stress amplitude of 530 MPa, no failure took place, even at 10^9 cycles. However, in some of the tests, the specimen failed beyond 10^7 load cycles. They also showed that during VHCF, the initiation of cracks from the persistent slip bands does not depend on the corresponding stress value [65].

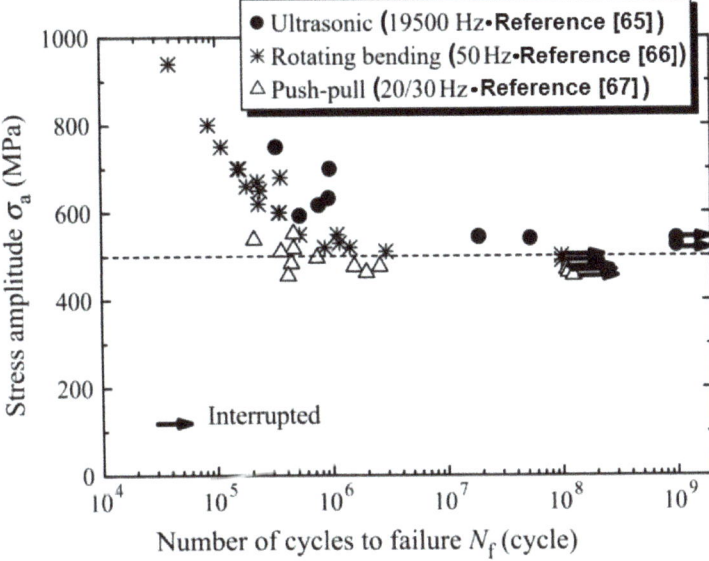

Figure 10. Very high cycle fatigue (VHCF) data of Inconel 718 by conventional and ultrasonic fatigue testing [65] (with permission from Elsevier, 2020).

The authors further studied and compared the effect of ultrasonic frequency on fatigue strength (rotating bending: 50 Hz by Kawagoishi et al. [66], and in push-pull: 20/30 Hz fatigue by Korth et al. [67]). The result showed that ultrasonic and rotating bending fatigue behave similarly, but there was a change in the chemical composition of the tested push-pull specimens. These results are similar to others showing longer lives for various metals in the VHCF regime with unexpected degenerative effects, and exceptions always exist [68].

3.5. Ti Alloys

Ti alloys are widely employed in aerospace industries where VHCF is common. In the VHCF range, the fatigue behavior of Ti alloys is similar to that of steels [64]. Bathias et al. tested the most widely employed Ti-6Al-4V alloy for fatigue at 20 kHz and R = −1. They revealed that the VHCF fatigue behavior of Ti-6Al-4V is better than conventional fatigue testing results at lower frequencies [3]. The specimen also did not tear out with increased cycles in the VHCF range. Yan et al. [69] studied the VHCF of Ti-6Al-4V and realized no fatigue limit, even after 10^9 cycles. The cracks initiated mostly on the surface and beneath the surface at the VHCF regime. Brittle and ductile fractures were observed during ultrasonic testing. During VHCF, the fatigue cracks initiated at heterogeneities like platelets. In Ti-6Al-4V, the duplex structure consisting of primary α-platelets, the crack initiated from these platelets.

Recently, Pan et al. studied the fatigue failure of gradient structured Ti-6Al-4V alloy, as shown in Figure 11 [70]. The gradient structure of the Ti alloy was obtained by the pretorsion experiment in the study theory. The results indicated that gradient Ti alloy showed better performance in LCF and HCF, but failed in the VHCF range. Thus, the gradient structure does not enhance the HCF strength of the Ti alloy. Their observations suggested that the VHCF is essential for the design of structural and long-service-life Ti components.

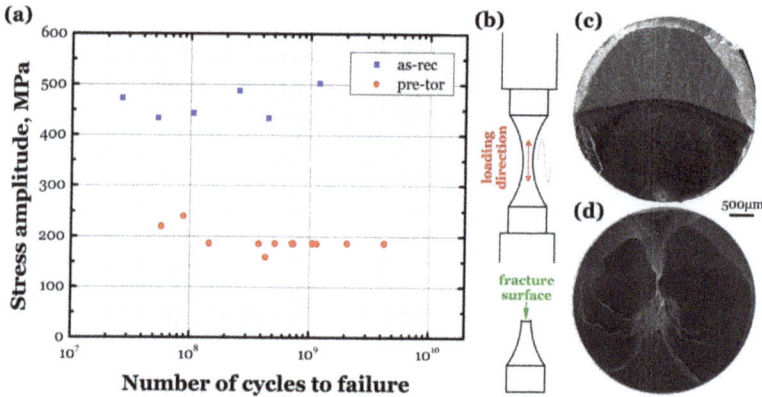

Figure 11. (a) S-N diagram at R = −1 for as-received and pre-torsioned (gradient structured) Ti alloy, (b) loading axis and fracture surface, (c) surface crack initiation in as-received Ti-6Al-4V, and (d) internal crack initiation in gradient structured Ti alloy, σ_a = 187 MPa, N_f = 2.07 × 10^9 [70].

The failure occurred between 2.64 × 10^7 and 1.19 × 10^9 cycles at σ_a = 434–503 MPa, and gradient structured alloy failed in the range of 5.78 × 10^7–4.23 × 10^9 cycles with a lower strength σ_a = 187 MPa. The failure modes showed surface crack initiation for as-received, and internal crack initiation for pre-torsioned specimens [70]. This type of fracture morphology is consistent with the fracture surfaces of other Ti alloys tested under the VHCF regime.

3.6. Cast Iron and Steels

Cast irons are cheap materials with good ductility, strength, and wear resistance. Spheroid graphite cast iron is most widely used in auto parts where the life expectancy of parts (suspension rods, gears, shafts, etc.) exceeds 10^9 cycles. The corresponding S-N diagram and fracture surfaces are shown in Figure 12. Wang et al. compared the VHCF of spheroidal cast iron at R = −1 and 0 with conventional fatigue tests [71]. They found that the failure of specimens continued to occur beyond 10^7 cycles without any fatigue limit; the effect of frequency was more pronounced for R = 0. The cracks were initiated at the surface for 10^7 load cycles and internal surface for greater than 10^7 load cycles. In another study, two different grades of ductile cast irons were examined [72].

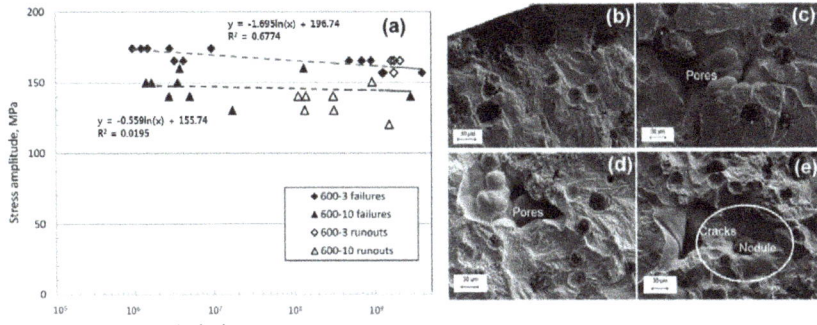

Figure 12. (a) S-N diagram of ferritic = pearlitic EN GJS-600-3 and ferritic EN GJS-600-10 iron grades, (b,c) fatigue fracture of EN GJS-600-3 specimen failure at 3.21×10^6 and 3.9×10^9 cycles, respectively, from scanning electron micrograph, (d,e) fatigue fracture of EN GJS-600-10 failure at 1.7×10^7 and 2.79×10^9 cycles, respectively [72] (with permission from John Wiley and Sons, 2020).

The results showed a higher fatigue strength (167 MPa) at 10^8 cycles of the EN-GJS-600-3 grade than that of the ferritic EN-GJS-600-10 grade (142 MPa). The microstructural observations showed that the presence of pores affected fatigue strength significantly. In some cases, cracks were nucleated at the nodule sites and propagated to ferrite island [72].

In another study, the authors tested low-carbon ferritic steel at R = −1. They showed that the fatigue strength was merely decreased by 25 MPa between 10^6 and 10^9 cycles without any fatigue limit. Figure 13a shows the S-N diagram of low-carbon steel at 20 kHz and R = 0.1 [73]. However, a significant difference of 200 MPa fatigue strength was observed between 10^6 and 10^9 cycles for 17-4PH martensitic stainless steel. In contrast, the S-N diagram of spring steels was asymptotic in the VHCF regime until 10^{10} cycles [74].

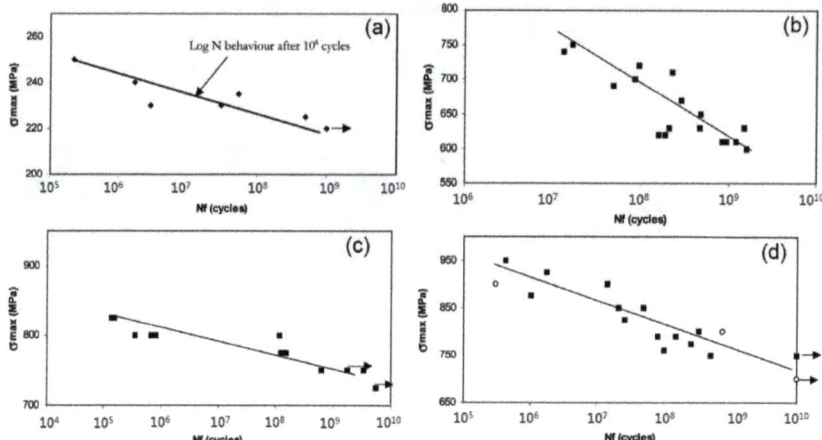

Figure 13. S-N diagram for (**a**) low carbon steel decreasing lifetime after 10^6 cycles, (**b**) 17-4PH martensitic stainless steel decreasing after 10^7 cycles, (**c,d**) 54SC6 and 54SC7 (respectively) spring steels until 10^{10} cycles at 20 kHz and R = −1 [73] (with permission from Elsevier, 2020).

Wang et al. noticed similar results in the investigation of ultra-high-strength steel springs made of Cr-V and Cr-Si steel at 20 kHz and R = −1. Cr-V steel showed high stability to fatigue in the VHCF range, while Cr-Si steel showed a drastic reduction in fatigue strength by 170 MPa beyond 10^9 cycles. At lower cycles, the crack initiated on the surface, while in the VHCF range, the cracks are present beneath the surface [75].

Sohar et al. studied the VHCF behavior of AISI D2 cold-worked steel at 20 kHz and R = −1. They found the crack initiation sites at primary carbides in the steel matrix and surface. These carbides were fractured during the fatigue and are responsible for a dramatic decrease in fatigue strength to 300 MPa in VHCF regime [76].

Sakai et al. investigated the VHCF properties of high C and high Cr-bearing steel (JIS: SUJ2) at 50 Hz and R = −1. The quenching of the specimens was done at 1108 K/40 min in oil. Air tempering of the specimens was also done at 453 K/120 min, followed by cooling. The fracture initiated on the surface and also inside the surface [77]. Besides, a circular region defined as the fish-eye was observed at the inclusion site in the air tempered sample.

Wang et al. also observed the fish-eye crack in his investigation on Cr-Fe-rich high-strength martensitic steel [78]. They observed the formation of the fish-eye at the site of the inclusions, as shown in Figure 14a–d. All these results are consistent with the reports on the VHCF study. The fish-eye is the most characteristic feature in the VHCF fracture formed by the change in the rate of crack propagation in the specimen at the discontinuity site.

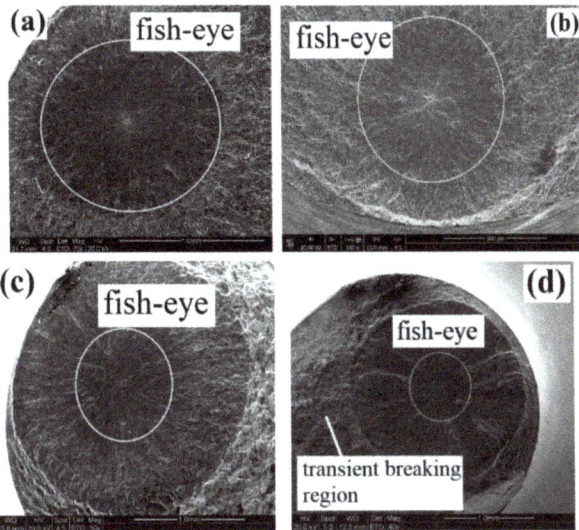

Figure 14. (a–d) The fish-eye region in high-Cr martensitic steel after VHCF [78].

Oh et al. investigated the influence of laser irradiation and vibration peening on the ultrasonic fatigue behavior of AISI4140 alloy at 20 kHz and R = −1 [79]. They found that laser irradiation increased the fatigue strength of AISI4140 when the temperature is around 700 °C or more, obtaining the highest fatigue strength at 800 °C. The laser irradiation provided sufficient energy to complete the transformation of austenite to martensite, which enhanced the fatigue behavior as compared to peened specimens.

3.7. Inference on VHCF of Engineering Materials

Most of the engineering metals and alloys do not exhibit fatigue limit; they even show a continuously reduced strength beyond 10^9 cycles. This behavior varies for different materials, and various factors in this process are not clearly understood in the literature. The only difference is that lower fatigue strength is obtained at 10^9 over 10^6 cycles in the VHCF regime. Therefore, the conventional fatigue limit of 10^6 cycles does not lie in the VHCF range.

Fortunately, the reason for fatigue failure can be discerned by the analysis of fracture surfaces. The weakest sites of failure can be identified as microstructural defects, inclusions, pores, inhomogeneity, platelets, and abnormal grain growth, as seen in Ti alloys or secondary reinforcements in the case of composites. The formation of slip bands is a major weak site in ductile metals like Cu or Al concerning VHCF. Amongst steels, the fatigue initiation sites are inclusions, carbides, pore, or slags segregations. The advanced high entropy alloys discovered recently have shown good fatigue behavior but there are limited investigations on the VHCF of these alloys, which is beyond the scope of this review [80–83].

4. Duplex S-N Curve

For materials undergoing VHCF failure, the conventional S-N diagram is modified as a duplex or multistage S-N curves, as shown in Figure 15 [84]. The existence of fatigue limits in the VHCF range is ambiguous, as shown in Figure 15. One has to distinguish between the surface and internal cracks based on the experimental evidence such as from fracture surfaces. A gradual shift from surface stress to the failures initiating the internal defects with higher cycles, leading to the development of "fish-eye", has been shown to occur [85,86].

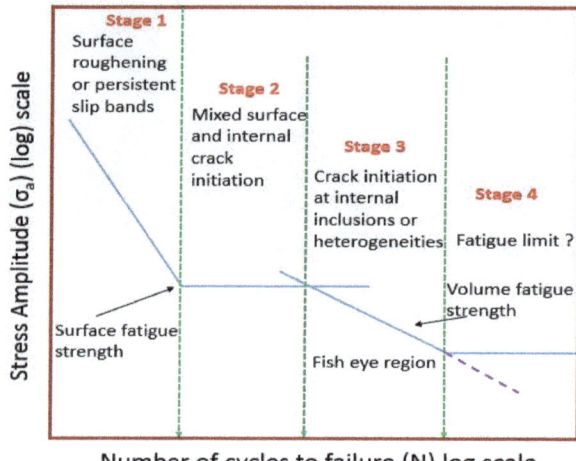

Figure 15. Duplex S-N diagram for various materials in different life cycles [87] (with permission from Elsevier, 2020).

4.1. Stages in Very High Cycle Fatigue

The fatigue fracture in the VHCF range can be split into multiple stages. (1) Stage 1 describes an LCF region where failure occurs from the surface and depicts surface fatigue strength. (2) Stage 2 describes an HCF region. Here, the stress required for the initiation of the persistent band is smaller than the threshold stress. (3) Stage 3 shows the crack initiation at internal defects and crack propagation within and outside the fish-eye and depicts the volume fatigue strength. Stages 2 and 3 constitute the conventional S-N diagram. As already discussed, the fish-eye is a characteristic of VHCF, which is a circular zone with an area of 0.5–1 mm diameter surrounding the origin site as internal circular crack propagation [88].

Several researchers have shown that fatigue life for crack initiation is greater than 90% across 10^6–10^7 cycles for steels, Ti, Al, and Ni alloys. This value is higher for VHCF—up to 99%. Paris et al. reported that the number of cycles in crack propagation life is only a small fraction of total fatigue life [19,89,90]. Therefore, understanding of causes of fatigue nucleation in a lifetime for materials under certain conditions is required. (4) Stage 4 represents the growth of internal microcracks and their propagation, which is slower than the surface microcracks. This may be due to their partially reversible slip formation. The presence of this region has been debated in several reports. Several reports discussed the non-existence of Stage 4 [40,41,49,50,64]; however, there are exceptions as well [85].

4.2. Single Phase and Multi-Phase Materials

Single-phase (type I) and multi-phase materials (type II) have a distinct impact on the VHCF behavior, as shown in various reports [84,91,92]. It has been seen that for typical type II materials, for example, high-strength steel, Al, Ti, and Mg alloys [85,93–95], the crack initiation in the VHCF regime occurs at the foreign inclusions inside, which acquire a fish-eye shaped structure [96–100]. On the contrary, featureless crack initiation sites are also noticed without any fish-eye, showing no visible effect of inclusions in aluminium alloys [101]. Such exceptions in the VHCF regime are unclear. In this regard, Davidson [102] explained that the microplastic deformation is mostly concentrated along largest grains termed as "supergrains". In contrast, Knobbe et al. [103] observed that in dual phase steel, surrounding phases or grains also contribute to the crack initiation and propagation.

Mughrabi [84,104] et al. proposed that VHCF failure begins at the surface due to irreversible or partially reversible deformed grains at surface for Type I materials. Continuous loading in the VHCF regime causes rough surface and persistent slip band formation. Weidner et al. [105] demonstrated the formation of such slip bands in Cu in the VHCF regime. In addition, the stress amplitudes for crack propagation might be greater than the threshold for slip band formation [106,107].

The S-N diagram of type II materials has a multistage fatigue life. The failure of a 'classical' fatigue limit beyond 10^7 cycles is correlated to the crack initiation sites and heterogeneities in the microstructure [59,85,87]. In contrast, type I materials do not necessarily exhibit multistage S-N diagram. The crack initiation occurs on account of localized and not fully reversible plastic deformation at the surface. The localized formation of persistent slip bands during VHCF regime might be detected without failure of the material because of the lower stresses for crack growth [108,109].

4.3. Origins of Very High Cycle Fatigue Failures

As discussed, VHCF cracks generally initiate from the microstructural defects such as inclusion in steels, or pores in powder metallurgy processed materials. Large carbides or carbide clusters can also cause crack initiations in cast iron and some steels. In other materials such as Ti alloys, where there is no inclusion or pores, the crack originates from the other heterogeneities like platelets of primary alpha phase, abnormal grains, or perlite colonies. These microstructural discontinuities serve as stress concentration raisers, even at smaller loads for fatigue initiation.

In the VHCF range, the failure is mostly from the internal cracks due to a higher possibility of such defects present in the bulk rather than on the surface. Among soft materials (Cu or Al), VHCF begins from the surface roughening due to the formation of slip bands, even after several cycles. The formation of such persistent slip bands leads to immediate cracking and failure [59,60]. In cast Al-Si alloys, the VHCF that begins at the band is due to the presence of a high amount of pores on the specimen surface. It is to be noted that research on the VHCF behavior of materials is now gripping the research community but the major fatigue mechanisms are still under debate and need to be investigated in depth.

4.4. Effect of Various Factors on Very High Cycle Fatigue

Tschegg et al. performed a series of VHCF investigations on metallic materials at various stress ratios (R = −1 to +0.5). They observed an R-dependence of the fatigue crack growth rate for chromium steel [110]. They also investigated the effect of environment (air, humidity and dry air) on the fatigue crack growth of VHCF of AA2024-T3. The fatigue growth rate was found to lie in the order of humid air < dry air < vacuum [111]. Naito et al. studied the effect of carburization on VHCF behavior of Cr-Mo steel [17]. They found that dual-stage S-N property was dominating for carburized specimens, while a smooth S-N diagram was evident for electropolished specimens. Nakamura et al. [112] investigated the VHCF tests on Ti-6Al-4V at room atmosphere and in vacuum. Their results indicated that fatigue life in the vacuum was longer as compared to that in room atmosphere at higher stress (>860 MPa), but no difference was noticed at lower stress levels (<860 MPa) when internal fracture occurred. Nishijima et al. investigated the effect of temperature for SCMV2 steel [85]. It was shown that the S-N diagram had a clear fatigue limit at room temperature and 200°C. However, at elevated temperatures, the S-N diagram had another curve for internally initiated fracture beyond $N > 10^7$. There are other reports on environmental effects but the authors chose some selected ones [113,114].

Sakai et al. studied the effect of aging treatment on VHCF behavior of maraging steels in rotating bending [115]. They found that the fatigue life of the aged specimen is higher by 10–100 times than without aging. They explained that this is due to the result of precipitation strengthening in the aged specimen. However, there was no visible effect of aging on the fatigue limit (600–625 MPa). Therefore, great attention is needed in real life applications of maraging steels in VHCF regime.

4.5. Failure Mechanisms until "Conventional Fatigue Limit" and Beyond

The failure mechanisms beyond the conventional fatigue limit are associated with very low crack growth rates. In microcrystalline engineering materials, a number of heterogeneities exist in the material (inclusion, precipitates and grain boundaries) which can initiate the cracks. It is inferred that this is most destructive situation for an engineering plant designed for more than 10^6 cycles where low cyclic stresses at a high mean stress level with or without the aforementioned fracture mechanisms [116,117]. The cracks may penetrate below the surface and grow in size compared to the non-propagating cracks. There are certain reports which demonstrate multiple fracture failure (surface and interior failures, fish-eye-induced failures, granular-area-fish-eye-induced failure, and/or optically dark-area-fish-eye failures) [116–118]. New methods and models need to be consistently designed for 10^6–10^{12} cycles other than current ultrasonic fatigue testing to more precisely determine the cause of multiple fracture mechanisms.

5. Decreasing Fatigue Strength in Very High Cycle Fatigue Regime

A drastic decrease in fatigue strength has been reported in the past for metallic materials [119–123]. It is noteworthy that traditional servo-hydraulic fatigue equipment, resonators, and shakers operate up to 400 Hz compared to ultrasonic fatigue equipment, which allow frequencies up to 100 times larger (40 kHz). Such a high frequency makes the VHCF test possible in the giga cycles (10^9) up to 10^{12} within a short time [3,19,73,124–126]. On the other hand, ultrasonic fatigue testing has some demerits, for instance, small specimens (d < 10 mm) are needed. Cooling or self-heating of the specimen must be taken into account. A direct correlation of data from low frequency testing and from ultrasonic fatigue testing is not always feasible [126,127].

6. Future Prospects

The conventional fatigue testing can be performed with the usual frequencies at high cycles by developing high-efficiency testing machines in different loading conditions. Ultrasonic fatigue testing is a useful tool to study the fatigue life in actual ultrasonic loading and designing of new materials for mankind. The fatigue behaviors of materials under usual frequencies (<100 Hz) and ultrasonic frequencies should be carefully examined under nominal frequencies like 20–50 kHz. For materials that work under extreme conditions, we need to develop a standard fatigue procedure and analytical models in the near future, for the sake of comparison of fatigue test data at different load cycle regimes. The previous literature indicates that VHCF behavior of engineering materials is still not completely understood. The multiple sources of failure in VHCF require advanced methods or equipment, not often using current ultrasonic-based fatigue testing systems. Ultrasonic fatigue systems do not consider time-dependent effects, yet they are the only feasible means to evaluate the VHCF mechanisms. Therefore, there is a great need to develop a general model governing all fatigue regimes (including variations in loading forms, such as torsion and blending) which can be applied for fatigue life prediction of materials.

7. Concluding Remarks

Owing to the developments in high-strength materials, it has become necessary to study the fatigue of these materials beyond the regular load cycle regime. The material research developments in the VHCF regime are now rising worldwide. Among advanced HEAs, limited studies on VHCF exist. Based on the previous data of conventional alloys, the study and understanding of fatigue behavior, not only at low cycles but also at VHCF, is in demand for novel applications of HEAs. The fatigue behavior at elevated temperatures and in corrosive media should also be examined for the new generation of HEAs. Ultrasonic fatigue testing is an excellent option for the study of the VHCF behavior of engineering materials; it allows a range of loading cycles at various temperatures. It also allows the selection of environments for fatigue testing. Almost 99% of the fatigue life is needed for crack nucleation in the VHCF regime. Therefore, it is essential to understand the crack initiation mechanism and new tools to recognize the respective cause and conditions. Possible crack initiation sites could be metallurgical discontinuities and heterogeneities in the specimens: shrinkage, casting defects, inclusions, porosity, interfaces, unusual grain growth, platelets, formation of slip bands, etc. A standard database is also needed for testing VHCF.

Considering various influential VHCF factors and their relevance, it is difficult to develop a common model to understand VHCF for all steels or metals, or even a small domain like tool steels. Therefore, the best way to understand the fatigue behavior in the VHCF regime is only to test them using equipment and observe the changes at microstructural levels, individually. Consequently, we can figure out ways of improving the fatigue strength of a material. Although the experimental fatigue data are scarce in the literature, further verification and comparison can be made at VHCF. The fatigue of HEAs can be a key to overcoming and improving new materials and alloy designing for future applications.

Author Contributions: Conceptualization, A.S. and B.A.; methodology, M.C.O.; resources, B.A.; writing—original draft preparation, A.S. and M.C.O.; writing—review and editing, B.A.; visualization, M.C.O.; project administration, B.A.; funding acquisition, B.A. All authors have read and agreed to the published version of the manuscript.

Funding: This research was supported by Basic Science Research Program through the National Research Foundation of Korea (NRF) funded by the Ministry of Education (NRF-2018R1D1A1B07044481).

Conflicts of Interest: The authors declare no conflict of interest.

References

1. Dengel, D. Planung und Auswertung von Dauerschwingversuchen bei angestrebter statistischer Absicherung der Kennwerte. In *Verhalten von Stahl bei Schwingender Beanspruchug*; Stahleisen, M.B.H., Ed.; VDI-Verl: Düsseldorf, Germany, 1978; pp. 23–46.
2. Suresh, S. *Fatigue of Materials*; Cambridge University Press: Cambridge, UK, 2003.
3. Bathias, C.; Paris, P.C. *Gigacycle Fatigue in Mechanical Practice*; Marcel Dekker: New York, NY, USA, 2004.
4. Brainthwaite, F. On the fatigue and consequent fracture of metals. *Inst. Civil Eng. Minutes Proc.* **1854**, *13*, 463–474.
5. Moore, H.F.; Kommers, J.B. *The Fatigue of Metals*; McGrawHill Book Company, Inc.: New York, NY, USA, 1927.
6. Wöhler, A.Z. Versuche über die Festigkeit der Eisenbahnwagenachsen. *Z. Bauwes.* **1860**, *10*, 160–161.
7. Gough, H.J. *The Fatigue of Metals*; Ernest Benn Ltd.: London, UK, 1926.
8. Bhat, S.; Patibandla, R. Metal fatigue and basic theoretical models: A review. In *Alloy Steel-Properties and Use*; Morales, E.V., Ed.; InTech: Rijeka, Croatia, 2011.
9. Basquin, O.H. The exponential law of endurance tests. *Proc. Am. Soc. Test. Mater.* **1910**, *10*, 625–630.
10. Langer, B.F. Design of pressure vessels for low-cycle fatigue. *J. Basic Eng.* **1962**, *84*, 389–399. [CrossRef]
11. Kurek, A.; Koziarska, J.; Łagoda, T. Strain characteristics of non-ferrous metals obtained on the basic of diffeerent loads. In Proceedings of the MATEC Web of Conferences, 12th International Fatigue Congress, Poitiers, France, 27 May–10 June 2018; p. 15005.
12. Kandil, F.A. The determination of uncertainties in low cycle fatigue testing. *Stand. Meas. Test. Proj.* **2000**, *1*, 1–28.
13. Łagoda, T. Energy models for fatigue life estimation under uniaxial random loading. Part II: Verification of the model. *Int. J. Fatigue* **2001**, *23*, 481–489. [CrossRef]
14. Toasa Caiza, P.D.; Ummenhofer, T.; Correia, J.A.F.O.; Jesus, A.D. Applying the Weibull and Stüssi methods that derive reliable Wöhler curves to historical German bridges. *Pract. Period. Struct. Des. Constr.* **2020**, *25*, 04020029. [CrossRef]
15. Barbosa, J.F.; Correia, J.A.F.O.; Junior, R.C.S.F.; Zhu, S.P.; De jesus, A.M.P. Probabilistic S-N fields based on statistical distributions applied to metallic and composite materials: State of the art. *Adv. Mech. Eng.* **2019**, *11*, 1–22. [CrossRef]
16. Correia, J.A.F.O.; Raposo, P.; Calvente, M.M.; Blason, S.; Lesiuk, G.; De Jesus, A.M.P.; Moreira, P.M.G.P.; Calcada, R.A.B.; Canteli, A.F. A generalization of the fatigue Kohout-Věchet model for several fatigue damage parameters. *Eng. Fract. Mech.* **2017**, *185*, 284–300. [CrossRef]
17. Naito, T.; Ueda, H.; Kikuchui, M. Fatigue behavior of carburized steel with internal oxides and nonmartensitic micro-structure near the surface. *Metall. Trans.* **1984**, *15A*, 1431–1436. [CrossRef]

18. Asami, K.; Sugiyama, Y. Fatigue strength of various surface hardened steels. *J. Heat. Treat. Technol. Assoc.* **1985**, *25*, 147–150.
19. Stanzl-Tschegg, S.E.; Mayer, H.R. Fatigue and fatigue crack growth of aluminium alloys at very high numbers of cycles. *Int. J. Fatigue* **2001**, *23*, 231–237. [CrossRef]
20. Mayer, H.; Schuller, R.; Fitzka, M. Fatigue of 2024-T351 aluminium alloy at different load ratios up to 1010 cycles. *Int. J. Fatigue* **2013**, *57*, 113–119. [CrossRef]
21. Morrissey, R.J.; Nicholas, T. Fatigue strength of Ti-6Al-4V at very long lives. *Int. J. Fatigue* **2005**, *27*, 1608–1612. [CrossRef]
22. Szczepanski, C.J.; Jha, S.K.; Larsen, J.M.; Jones, J.W. Microstructural influences on very-high-cycle fatigue crack initiation in Ti-6246. *Metall. Mater. Trans. A* **2008**, *39*, 2841–2851. [CrossRef]
23. Furuya, Y.; Takeuchi, E. Gigacycle fatigue properties of Ti-6Al-4V alloy under tensile mean stress. *Mater. Sci. Eng. A* **2014**, *598*, 135–140. [CrossRef]
24. Liu, X.; Sun, C.; Hong, Y. Effects of stress ratio on high-cycle and very-high-cycle fatigue behavior of a Ti-6Al-4V alloy. *Mater. Sci. Eng. A* **2015**, *622*, 228–235. [CrossRef]
25. Li, S.X. Effects of inclusions on very high cycle fatigue properties of high strength steels. *Int. Mater. Rev.* **2012**, *57*, 92–114. [CrossRef]
26. Zimmermann, M. Diversity of damage evolution during cyclic loading at very high numbers of cycles. *Int. Mater. Rev.* **2012**, *57*, 73–91. [CrossRef]
27. Bathias, C. *Fatigue Limit in Metals*; Focus Series; John Wiley and Sons, Inc.: Hoboken, NJ, USA, 2003.
28. Liu, H.; Wang, H.; Huang, Z.; Wang, Q.; Chen, Q. Comparative study of very high cycle tensile and torsional fatigue in TC17 titanium alloy. *Int. J. Fatigue* **2020**, *139*, 105720. [CrossRef]
29. Kim, Y.; Hwang, W. High-cycle, low-cycle, extremely low-cycle fatigue and monotonic fracture behaviors of low-carbon steel and its welded joint. *Materials* **2019**, *12*, 4111. [CrossRef] [PubMed]
30. Kanazawa, K.; Yamaguchi, K.; Nishijima, S. Mapping of low cycle fatigue mechanisms at elevated temperatures for an austenitic stainless steel. *ASTM Spec. Tech. Publ.* **1988**, *942*, 519–530.
31. Murakami, Y.; Miller, K.J. What is fatigue damage? A viewpoint from the observation of a low cycle fatigue process. *Int. J. Fatigue* **2005**, *27*, 991–1005. [CrossRef]
32. Campbell, R.D. Creep/fatigue interaction correlation for 304 stainless steel subjected to strain-controlled cycling with hold times at peak strain. *J. Eng. Ind.* **1971**, *93*, 887–892. [CrossRef]
33. Coffin, L.F.M. Corrosion fatigue. *Natl. Assoc. Corros. Eng.* **1972**, *2*, 590–600.
34. Wareing, J.; Tomkins, B.; Sumner, G. Fatigue at elevated temperatures. *Am. Soc. Test. Mater. ASTM STP* **1973**, *520*, 123–138.
35. Yamaguchi, K.; Kanazawa, K. Influence of grain size on the low-cycle fatigue lives of austenitic stainless steels at high temperatures. *Metall. Trans. A* **1980**, *10*, 1691–1699. [CrossRef]
36. Beer, F.P.; Johnston, E.R. *Mechanics of Materials*; McGraw-Hill: New York, NY, USA, 1992.
37. Tomaszewski, T. Statistical size effect in fatigue properties for mini-specimens. *Materials* **2020**, *13*, 2384. [CrossRef] [PubMed]
38. Forsyth, P.J.E.; Stubbington, C.A.; Clark, D. Cleavage facets observed on fatigue-facture surfaces in an aluminum alloy. *J. Inst. Met.* **1962**, *90*, 238–239.
39. Zhao, P.; Wang, X.R.; Yan, E.; Misra, R.D.K.; Du, C.M.; Du, F. The influence of inclusion factors on ultra-high cyclic deformation of a dual phase steel. *Mater. Sci. Eng. A* **2019**, *754*, 275–281. [CrossRef]
40. Martina, Z. Very high cycle fatigue. In *Handbook of Mechanics of Materials*; Hsueh, C.-H., Schmauder, S., Chen, C.S., Chawla, K.K., Chawla, N., Chen, W., Kagawa, Y., Eds.; Springer: Singapore, 2018.
41. Kuhn, H.; Medlin, D. Mechanical Testing and Evaluation. In *ASM Handbook*; ASM International: Materials Park, OH, USA, 2000; Volume 8.
42. Mayer, H. Recent developments in ultrasonic fatigue. *Fatigue Fract. Eng. Mater. Struct.* **2016**, *39*, 3–29. [CrossRef]
43. Stanzl-Tschegg, S.E.; Mayer, H.R.; Tschegg, E.K. High frequency method for torsion fatigue testing. *Ultrasonics* **1993**, *31*, 275–280. [CrossRef]
44. Mayer, H. Ultrasonic torsion and tension-compression fatigue testing: Measuring principles and investigations on 2024-T351 aluminium alloy. *Int. J. Fatigue* **2006**, *28*, 1446–1455. [CrossRef]

45. Nikitin, A.; Bathias, C.; Palin-Luc, T. A new piezoelectric fatigue testing machine in pure torsion for ultrasonic gigacycle fatigue tests: Application to forged and extruded titanium alloys. *Fatigue Fract. Eng. Mater. Struct.* **2015**, *38*, 1294–1304. [CrossRef]
46. Wagner, D.; Cavalieri, F.J.; Bathias, C.; Ranc, N. Ultrasonic fatigue tests at high temperature on an austenitic steel. *Propuls. Power Res.* **2012**, *1*, 29–35. [CrossRef]
47. Palin-Luc, T.; Perez-Mora, R.; Bathias, C.; Dominguez, G.; Paris, P.C.; Arana, J.-L. Fatigue crack initiation and growth on a steel in the very high cycle regime with sea water corrosion. *Eng. Fract. Mech.* **2010**, *77*, 1953–1962. [CrossRef]
48. Perez-Mora, R.; Palin-Luc, T.; Bathias, C.; Paris, P.C. Very high cycle fatigue of a high strength steel under sea water corrosion: A strong corrosion and mechanical damage coupling. *Int. J. Fatigue* **2015**, *74*, 156–165. [CrossRef]
49. Marines-Garcia, I.; Paris, P.C.; Tada, H.; Bathias, C.; Lados, D. Fatigue crack growth from small to large cracks on very high cycle fatigue with fish-eye failures. *Eng. Fract. Mech.* **2008**, *75*, 1657–1665. [CrossRef]
50. Hailong, D.; Wei, L.; Tatsuo, S.; Zhenduo, S. Very high cycle fatigue failure analysis and life prediction of Cr-Ni-W gear steel based on crack initiation and growth behaviors. *Materials* **2015**, *8*, 8338–8354.
51. Wang, Q.Y.; Li, T.; Zeng, X.Z. Gigacycle fatigue fehavior of high strength aluminum alloys. *Procedia Eng.* **2010**, *2*, 65–70. [CrossRef]
52. Lee, B.H.; Park, S.W.; Hyun, S.K.; Cho, I.S.; Kim, K.T. Mechanical properties and very high cycle fatigue behavior of peak-aged AA7021 alloy. *Metals* **2018**, *8*, 1023. [CrossRef]
53. Oh, K.K.; Kim, Y.W.; Kim, J.H. High cycle fatigue characteristics of aluminum alloy by shot peening. *Adv. Mater. Res.* **2015**, *1110*, 142–147. [CrossRef]
54. Koutiri, I.; Bellett, D.; Morel, F.; Augustins, L.; Adrien, J. High cycle fatigue damage mechanisms in cast aluminium subject to complex loads. *Int. J. Fatigue* **2013**, *47*, 44–57. [CrossRef]
55. Yang, F.; Yin, S.M.; Li, S.X.; Zhang, Z.F. Crack initiation mechanism of extruded AZ31 magnesium alloy in the very high cycle fatigue regime. *Mater. Sci. Eng. A* **2008**, *491*, 131–136. [CrossRef]
56. Karr, U.; Stich, A.; Mayer, H. Very high cycle fatigue of wrought magnesium alloy AZ61. *Procedia Struct. Integrity* **2016**, *2*, 1047–1054. [CrossRef]
57. Bhuiyan, M.S.; Mutoh, Y.; Murai, T.; Iwakam, S. Corrosion fatigue behavior of extruded magnesium alloy AZ80-T5 in a 5% NaCl environment. *Eng. Fract. Mech.* **2010**, *77*, 1567–1576. [CrossRef]
58. Nascimento, L.; Yi, S.; Bohlen, J.; Fuskova, L.; Letzig, D.; Kainer, K.U. High cycle fatigue behaviour of magnesium alloys. *Procedia Eng.* **2010**, *2*, 743–750. [CrossRef]
59. Mughrabi, H. Specific features and mechanisms of fatigue in the ultrahigh-cycle regime. *Int. J. Fatigue* **2006**, *28*, 1501–1508. [CrossRef]
60. Mughrabi, H.; Hoppel, H.W.; Kautz, M. Fatigue and microstructure of ultrafine-grained metals produced by severe plastic deformation. *Scr. Mater.* **2004**, *51*, 807–812. [CrossRef]
61. Kunz, L.; Lukas, P.; Svoboda, M. Fatigue strength, microstructural stability and strain localization in ultrafine-grained copper. *Mater. Sci. Eng. A* **2006**, *424*, 97–104. [CrossRef]
62. Agnew, S.R.; Vinogradov, A.Y.; Hashimoto, S.; Weertman, J.T. Fatigue crack growth and related microstructure evolution in ultrafine grain copper processed by ECAP. *Mater. Trans.* **2012**, *53*, 101–108.
63. Vinogradov, A.; Hashimoto, S. Multiscale phenomena in fatigue of Ultra-fine grain materials—An Overview. *Mater. Trans.* **2001**, *42*, 74–84. [CrossRef]
64. Bathias, C. There is no infinite fatigue life in metallic materials. *Fatigue Fract. Eng. Mater. Struct.* **1999**, *22*, 559–565. [CrossRef]
65. Chen, Q.N.; Kawagoishi, Q.Y.; Wang, N.; Yan, T.; Ono, G.; Hashiguchi, G. Small crack behavior and fracture of nickel-based superalloy under ultrasonic fatigue. *Int. J. Fatigue* **2005**, *27*, 1227–1232. [CrossRef]
66. Kawagoishi, N.; Chen, Q.; Nisitani, H. Fatigue strength of Inconel 718 at elevated temperatures. *Fatigue Fract. Eng. Mater. Struct.* **2000**, *23*, 209–217. [CrossRef]
67. Korth, G.E.; Smolik, G.R. *Status Report of Physical and Mechanical Test of Alloy 718*; Report TREE-1254; EG&G Idaho, Inc.: Idaho Falls, ID, USA, 1978.
68. Willertz, L.E. Ultrasonic fatigue. *Int. Met. Rev.* **1980**, *25*, 65–78. [CrossRef]
69. Yan, N.; Zhu, X.; Han, D.; Liu, F.; Yu, Y. Very high cycle fatigue behavior of Ti-6Al-4V alloy. In Proceedings of the 4th Annual International Conference on Material Engineering and Application (ICMEA 2017), Wuhan, China, 15–17 December 2017.

70. Pan, X.; Qian, G.; Wu, S.; Fu, Y.; Hong, Y. Internal crack characteristics in very-high-cycle fatigue of a gradient structured titanium alloy. *Sci. Rep.* **2020**, *10*, 4742. [CrossRef]
71. Wang, Q.Y.; Bathias, C. Fatigue characterization of a spheroidal graphite cast iron under ultrasonic loading. *J. Mater. Sci.* **2004**, *39*, 687–689. [CrossRef]
72. Bergstrom, J.; Burman, C.; Svensson, J.; Jansson, A.; Ivansson, C.; Zhou, J.; Valizadeh, S. Very high cycle fatigue of two ductile iron grades. *Steel Res. Int.* **2016**, *87*, 614–621. [CrossRef]
73. Marines, I.; Bin, X.; Bathias, C. An understanding of very high cycle fatigue of metals. *Int. J. Fatigue* **2003**, *25*, 1101–1107. [CrossRef]
74. Bathias, C.; Drouillac, L.; Francois, P.L. How and why the fatigue S-N curve does not approach a horizontal asymptote. *Int. J. Fatigue* **2001**, *23*, 143–151. [CrossRef]
75. Wang, Q.Y.; Berard, J.Y.; Rathery, S.; Bathias, C. High-cycle fatigue crack initiation and propagation behaviour of high-strength spring steel wires. *Fatigue Fract. Eng. Mater. Struct.* **1999**, *22*, 673–677. [CrossRef]
76. Sohar, C.; Betzwar-Kotas, A.; Gierl, C.; Weiss, B.; Danninger, H. Gigacycle fatigue behaviour of a high chromium alloyed cold work tool steel. *Int. J. Fatigue* **2008**, *30*, 1137–1149. [CrossRef]
77. Sakai, T.; Sato, Y.; Oguma, N. Characteristic S-N properties of high-carbon-chromium bearing steel under loading in long-life fatigue. *Fatigue Fract. Eng. Mater. Struct.* **2002**, *25*, 765–773. [CrossRef]
78. Wang, J.; Yang, Y.; Yu, J.; Wang, J.; Du, F.; Zhang, Y. Fatigue life evaluation considering fatigue reliability and fatigue crack for FV520B-I in VHCF regime based on fracture mechanics. *Metals* **2020**, *10*, 371. [CrossRef]
79. Oh, M.C.; Yeon, H.; Jeon, Y.; Ahn, B. Microstructural characterization of laser heat treated AISI 4140 steel with improved fatigue behavior. *Arch. Metall. Mater.* **2015**, *60*, 1331–1334. [CrossRef]
80. Cantor, B.; Chang, I.T.H.; Knight, P.; Vincent, A.J.B. Microstructural development in equiatomic multicomponent alloys. *Mater. Sci. Eng. A* **2004**, *375–377*, 213–218. [CrossRef]
81. Zhang, Y.; Zuo, T.T.; Tang, Z.; Gao, M.C.; Dahmen, K.A.; Liaw, P.K. Microstructures and properties of high-entropy alloys. *Prog. Mater. Sci.* **2017**, *61*, 1–93. [CrossRef]
82. Sharma, A. High-Entropy Alloys for Micro- and Nanojoining Applications. In *Engineering Steels and High Entropy-Alloys*; Sharma, A., Ed.; Intechopen: Rijeka, Croatia, 2020.
83. Sharma, A.; Kumar, S.; Duriagina, Z. *Engineering Steels and High Entropy-Alloys*; IntechOpen Publishers: Rijeka, Croatia, 2020.
84. Mughrabi, H. On 'multi-stage' fatigue life diagrams and the relevant life-controlling mechanisms in ultrahigh-cycle fatigue. *Fatigue Fract. Eng. Mater. Sruct.* **2002**, *25*, 755–764. [CrossRef]
85. Nishijima, S.; Kanazawa, K. Stepwise S-N curve and fish-eye failure in gigacycle fatigue. *Fatigue Fract. Eng. Mater. Struct.* **1999**, *22*, 601–607. [CrossRef]
86. Tian, H.; Kirkham, M.J.; Jiang, L.; Yang, B.; Wang, G.; Liaw, P.K. A review of failure mechanisms of ultra high cycle fatigue in engineering materials. In Proceedings of the 4th Internationa Conference on Very High Cycle Fatigue, VHCF-4, Ann Arbor, MI, USA, 19–22 August 2007; pp. 437–444.
87. Pyttel, B.; Schwerdt, D.; Berger, C. Very high cycle fatigue—Is there a fatigue limit? *Int. J. Fatigue* **2011**, *33*, 49–58. [CrossRef]
88. Tridelloa, A.; Paolinoa, D.S.; Chiandussia, G.; Rossetto, M. VHCF strength decrement in large H13 steel specimens subjected to ESR process. *Procedia Struct. Integr.* **2016**, *2*, 1117–1124. [CrossRef]
89. Billaudeau, T.; Nadot, Y. Support for an environmental effect on fatigue mechanisms in the long life regime. *Int. J. Fatigue* **2004**, *26*, 839–847. [CrossRef]
90. Marines-Garcia, I.; Paris, P.C.; Tada, H.; Bathias, C. Fatigue crack growth from small to large cracks in gigacycle fatigue with fish-eye failures. In Proceedings of the 9th International Fatigue Congress, Atlanta, GA, USA, 14–19 May 2006.
91. Bayraktar, E.; Marines-Garcia, I.; Bathias, C. Failure mechanisms of automotive metallic alloys in very high cycle fatigue range. *Int. J. Fatigue* **2006**, *28*, 1521–1532. [CrossRef]
92. Tschegg, S.S.; Mughrabi, H.; Schönbauer, B. Life time and cyclic slip of copper in the VHCF regime. *Int. J. Fatigue* **2007**, *29*, 2050–2059.
93. Murakami, Y.; Nomoto, T.; Ueda, T. Factors influencing the mechanism of superlong fatigue failure in steels. *Fatigue Fract. Eng. Mater. Struct.* **1999**, *22*, 581–590. [CrossRef]
94. Zhu, K.; Jones, J.W.; Mayer, H.; Lasecki, J.V.; Allison, J.E. Effects of microstructure and temperature on fatigue behavior of E319-T7 cast aluminum alloy in very long life cycles. *Int. J. Fatigue* **2006**, *28*, 1566–1571. [CrossRef]

95. Berger, C.; Pyttel, B.; Trossmann, T. Very high cycle fatigue tests with smooth and notched specimens and screws made of light metal alloys. *Int. J. Fatigue* **2006**, *28*, 1640–1646. [CrossRef]
96. Awatani, J.; Katagiri, K.; Omura, A.; Shiraishi, T. Study of the fatigue limit of copper. *Metall. Trans. A* **1975**, *6*, 1029–1034. [CrossRef]
97. Nguyen, H.Q.; Gallimard, L.; Bathias, C. Numerical simulation of fish eye fatigue crack growth in very high cycle fatigue. *Eng. Fract. Mech.* **2015**, *135*, 81–93. [CrossRef]
98. Furuya, Y.; Matsuoka, S. Improvement of gigacycle fatigue properties by modified ausforming in 1600 and 2000 MPa-class low-alloy steel. *Met. Trans. A* **2002**, *33*, 3421. [CrossRef]
99. Mayer, H.; Haydn, W.; Schuller, R.; Issler, S.; Bacher-Höchst, M. Very high cycle fatigue properties of bainitic high carbon-chromium steel under variable amplitude conditions. *Int. J. Fatigue* **2009**, *31*, 242. [CrossRef]
100. Terent'ev, V.F. On the Problem of the Fatigue Limit of Metallic Materials. *Metal Sci. Heat Treat.* **2004**, *46*, 244–249. [CrossRef]
101. Pyttel, B.; Schwerdt, C.; Berger, C. Very high cycle fatigue behaviour of two different aluminium wrought alloys. In Proceedings of the 4th Internationa Conference on Very High Cycle Fatigue, VHCF-4, Ann Arbor, MI, USA, 19–22 August 2007; Allison, J.E., Jones, J.W., Larsen, J.M., Ritchie, R.O., Eds.; pp. 313–318.
102. Davidson, D.L. The effect of a cluster of similarly oriented grains (a supergrain) on fatigue crack initiation characteristics of clean material. In Proceedings of the 4th International Conference on Very High Cycle Fatigue, VHCF-4, Ann Arbor, MI, USA, 19–22 August 2007; Allison, J.E., Jones, J.W., Larsen, J.M., Ritchie, R.O., Eds.; pp. 23–28.
103. Knobbe, H.; Köster, P.; Krupp, U.; Christ, H.J.; Fritzen, C.P.; Anis Cherif, M.; Altenberger, I. Crack Initiation and propagation in a Stainless Duplex Steel during HCF and VHCF. In Proceedings of the 4th Internationa Conference on Very High Cycle Fatigue, VHCF-4, Ann Arbor, MI, USA, 19–22 August 2007; Allison, J.E., Jones, J.W., Larsen, J.M., Ritchie, R.O., Eds.; pp. 143–149.
104. Mughrabi, H. On the life-controlling microstructural fatigue mechanisms in ductile metals and alloys in the giga cycle regime. *Fatigue Fract. Eng. Mater. Struct.* **1999**, *22*, 633–641. [CrossRef]
105. Weidner, A.; Amberger, D.; Pyczak, F.; Schönbauer, B.; Tschegg, S.S.; Mughrabi, H. Fatigue damage in copper polycrystals subjected to ultrahigh-cycle fatigue below the PSB threshold. *Int. J. Fatigue* **2010**, *32*, 872–878. [CrossRef]
106. Mughrabi, H.; Tschegg, S.S. Fatigue damage evolution in ductile single-phase face-centered cubic metals in the UHCF-regime. In Proceedings of the 4th Internationa Conference on Very High Cycle Fatigue, VHCF-4, Ann Arbor, MI, USA, 19–22 August 2007; Allison, J.E., Jones, J.W., Larsen, J.M., Ritchie, R.O., Eds.; pp. 75–82.
107. Lukáš, P.; Klesnil, M.; Polák, J. High cycle fatigue life of metals. *Mater. Sci. Eng. A* **1974**, *15*, 239–245. [CrossRef]
108. Höppel, H.W.; Saitova, L.R.; Grieß, H.J.; Göken, M. Surface roughening and fatigue behaviour of pure aluminium with various grain sizes in the VHCF-regime. In Proceedings of the 4th Internationa Conference on Very High Cycle Fatigue, VHCF-4, Ann Arbor, MI, USA, 19–22 August 2007; Allison, J.E., Jones, J.W., Larsen, J.M., Ritchie, R.O., Eds.; pp. 59–66.
109. Höppel, H.W.; May, L.; Prell, M.; Göken, M. Influence of grain size and precipitation state on the fatigue lives and deformation mechanisms of CP aluminium and AA6082 in the VHCF regime. *Int. J. Fatigue* **2011**, *33*, 10–18. [CrossRef]
110. Stanzl, E.; Czegley, M.; Mayer, H.; Tschegg, E. Fatigue Crack Growth under Combined Mode I and Mode II Loading. In *Fracture Mechanics: Perspectives and Directions (Twentieth Symposium)*; Wei, R., Gangloff, R., Eds.; ASTM International: West Conshohocken, PA, USA, 1989; pp. 479–496.
111. Tschegg, S.S. Fracture mechanisms and fracture mechanics at ultrasonic frequencies. *Fatigue Fract. Eng. Mater. Struct.* **1999**, *22*, 567–579. [CrossRef]
112. Nakamura, T.; Koneko, M.; Kazami, S.; Noguchi, T. The Effect of High Vacuum Environment on Tensile Fatigue Properties of Ti-6Al-4V Alloy. *J. Soc. Mater. Sci. Jpn.* **2000**, *49*, 1148–1154. [CrossRef]
113. Petit, J.; Saraazin-Boudox, C. An overview on the influence of the atmosphere environment on ultra-high-cycle fatigue and ultra-slow fatigue crack propagation. *Int. J. Fatigue* **2006**, *28*, 1471–1478. [CrossRef]
114. Miura, N.; Takahashi, Y. High-cycle fatigue behavior of type 316 stainless steel at 288 °C including mean stress effect. *Int. J. Fatigue* **2006**, *28*, 1618–1625. [CrossRef]

115. Sakai, T.; Chen, Q.; Uchiyama, A.; Nakagawa, A.; Ohnaka, T. A study on ultra-long life fatigue characteristics of maraging steels with/without aging treatment in rotating bending. In Proceedings of the 4th Internationa Conference on Very High Cycle Fatigue, VHCF-4, Ann Arbor, MI, USA, 19–22 August 2007.
116. Shiozawa, K.; Lu, L. Internal fatigue failure mechanism of high strength steels in Gigacycle regime. *Key Eng. Mater.* **2018**, *378–379*, 65–68. [CrossRef]
117. Murakami, Y.; Nomoto, T.; Ueda, T.; Murakami, Y. On the mechanism of fatigue failure in the superlong life regime (N > 107 cycles). Part II: A fractographic investigation. *Fatigue Fract. Eng. Mater. Struct.* **2000**, *23*, 903–910. [CrossRef]
118. Hu, Y.; Sun, C.; Xie, J.; Hong, Y. Effects of loading frequency and loading type on high-cycle and very-high-cycle fatigue of a high-strength steel. *Materials* **2018**, *11*, 1456. [CrossRef] [PubMed]
119. Marines, I.; Dominguez, G.; Baudry, G.; Vittori, J.-F.; Rathery, S.; Doucet, J.-P.; Bathias, C. Ultrasonic fatigue tests on bearing steel AISI-SAE 52100 at frequency of 20 and 30 kHz. *Int. J. Fatigue* **2003**, *25*, 1037–1046. [CrossRef]
120. Mayer, H. Fatigue damage of low amplitude cycles under variable amplitude loading condition. In Proceedings of the 4th Internationa Conference on Very High Cycle Fatigue, VHCF-4, Ann Arbor, MI, USA, 19–22 August 2007.
121. Schwerdt, D.; Pyttel, B.; Berger, C.; Oechsner, M.; Kunz, U. Microstructure investigations on two different aluminum wrought alloys after very high cycle fatigue. *Int. J. Fatigue* **2014**, *60*, 28–33. [CrossRef]
122. Sakai, T. Review and prospects for current studies on very high cycle fatigue of metallic materials for machine structural use. *J. Solid Mech. Mater. Eng.* **2009**, *3*, 425–439. [CrossRef]
123. Sonsino, C.M. *Dauerfestigkeit—Eine Fiktion*; Konstruktion 4: Fraunhofer-Gesellschaft: Darmstadt, Germany, 2005.
124. Tschegg, S.S.; Mayer, H.; Tschegg, E.; Beste, A. In service loading of Al–Si11 aluminium cast alloy in the very high cycle regime. *Int. J. Fatigue* **1993**, *15*, 311–316.
125. Mayer, H.; Papakyriacou, M.; Pippan, R.; Tschegg, S.S. Influence of loading frequency on the high cycle fatigue properties of AlZnMgCu1.5 aluminium alloy. *Mater. Sci. Eng. A* **2001**, *314*, 48–54. [CrossRef]
126. Tsutsumi, N.; Murakami, Y.; Doquet, V. Effect of test frequency on fatigue strength of low carbon steel. *Fatigue Fract. Eng. Mater. Struct.* **2009**, *32*, 473–483. [CrossRef]
127. Tschegg, S.S. Ultrasonic fatigue. In Proceedings of the International Conference on Fatigue 96, Berlin, Germany, 6–10 May 1996; Volume III, pp. 1887–1898.

 © 2020 by the authors. Licensee MDPI, Basel, Switzerland. This article is an open access article distributed under the terms and conditions of the Creative Commons Attribution (CC BY) license (http://creativecommons.org/licenses/by/4.0/).

Article

Capturing and Micromechanical Analysis of the Crack-Branching Behavior in Welded Joints

Wenjie Wang [1], Jie Yang [1,*], Haofeng Chen [2] and Qianyu Yang [1]

[1] Shanghai Key Laboratory of Multiphase Flow and Heat Transfer in Power Engineering, School of Energy and Power Engineering, University of Shanghai for Science and Technology, Shanghai 200093, China; wenjiewang0426@163.com (W.W.); qyyang9908@163.com (Q.Y.)
[2] Department of Mechanical & Aerospace Engineering, University of Strathclyde, Glasgow G1 1XJ, UK; haofeng.chen@strath.ac.uk
* Correspondence: yangjie@usst.edu.cn; Tel.: +86-021-5527-2320

Received: 25 August 2020; Accepted: 27 September 2020; Published: 29 September 2020

Abstract: During the crack propagation process, the crack-branching behavior makes fracture more unpredictable. However, compared with the crack-branching behavior that occurs in brittle materials or ductile materials under dynamic loading, the branching behavior has been rarely reported in welded joints under quasi-static loading. Understanding the branching criterion or the mechanism governing the bifurcation of a crack in welded joints is still a challenge. In this work, three kinds of crack-branching models that reflect simplified welded joints were designed, and the aim of the present paper is to find and capture the crack-branching behavior in welded joints and to shed light on its branching mechanism. The results show that as long as there is another large enough propagation trend that is different from the original crack propagation direction, then crack-branching behavior occurs. A high strength mismatch that is induced by both the mechanical properties and dimensions of different regions is the key of crack branching in welded joints. Each crack branching is accompanied by three local high stress concentrations at the crack tip. Three pulling forces that are created by the three local high stress concentrations pull the crack, which propagates along with the directions of stress concentrations. Under the combined action of the three pulling forces, crack branching occurs, and two new cracks initiate from the middle of the pulling forces.

Keywords: crack branching behavior; micromechanical analysis; crack propagation path; welded joints; stress concentration

1. Introduction

Cracks are the main drivers of material failure [1,2]. In the crack propagation process, a crack may split into two or more branches. This crack-branching phenomenon usually occurs in concrete structures, brittle materials, and quasi-brittle materials under dynamic loading, and it makes the fracture become more unpredictable and has aroused a wide range of concerns. For concrete structures, Forquin [3] investigated the crack propagation behavior in concrete and rock-like materials under dynamic tensile loading by an optical correlation technique. Curbach et al. [4] discussed the crack velocity in concrete by an experimental investigation. Ožbolt et al. [5–7] studied the inertia on resistance, failure mode, and crack pattern of concrete loaded by higher loading rates. Zhang et al. [8] reviewed the development and the state of the art in dynamic testing techniques and dynamic mechanical behavior of rock materials. For the brittle and quasi-brittle materials, much research has been done. Most recently, Mecholsky et al. [9] studied the relationship between fractography, fractal analysis, and crack branching in brittle materials. Nakamura et al. [10] researched the effect of the stress field on crack branching in brittle material. Chen et al. [11] studied the influence of micro-modulus functions on peridynamics simulation of crack propagation and branching in brittle materials. Kou et al. [12] investigated the crack propagation and crack branching in brittle solids under dynamic loading. Li et al. [13] studied the underlying fracture trends and triggering on Mode-II crack branching and kinking for quasi-brittle solids. Bouchbinder et al. [14] studied the dynamics of branching instabilities in rapid fracture and offered predictions for the geometry of multiple branches. Boué et al. [15] investigated the source of the micro-branching instability and revealed the relationship between micro-branching and the oscillatory instabilities of rapid cracks. Karma et al. [16] researched the unsteady crack motion and branching in brittle fracture and shed light on the physics that control the speed of accelerating cracks and the characteristic branching instability.

In the meantime, different classes of models and methods were selected to study the crack-branching behavior in the crack propagation process. The cohesive region model describes the crack propagation process by considering a potential opening between two bulk elements, and it can capture some features of crack-branching patterns [17,18]. The extended finite element method (XFEM) was also selected to obtain crack branching by input additional branching criterion in the crack propagation algorithm [19,20]. The phase field model was the most widely used to research the crack-branching behavior in the dynamic crack propagation process. Henry [21] studied the dynamic branching instability under in-plane loading by a phase field model. Bleyer [22] investigated the crack-branching, speed-limiting, and velocity-toughening mechanism in the dynamic crack propagation process by a variational phase-field model. Hofacker and Miehe [23,24] described the evolution of complex crack patterns under dynamic loading by representative numerical examples. Karma et al. [25] introduced a phenomenological continuum model for the mode III dynamic fracture that is based on the phase-field methodology. Henry and Adda-Bedia [26] studied the crack branching in brittle material and established its relationship to the fractographic patterns by a phase-field model. In addition, Bobaru and Zhang [27] reviewed the peridynamic model for brittle fracture and investigated the crack-branching behavior in brittle homogeneous and isotropic materials.

A crack can branch for many reasons. Since an additional crack was generated in the branching event, the energy release rate was proposed as a crack-branching criterion [28]. Another important crack-branching criterion is crack-tip velocity. There exists a critical value of crack-tip velocity, and crack branching occurs at the critical value [29]. Nevertheless, in the inelastic nonlocal continuum model, such as a phase-field model, the crack-branching behavior can be captured naturally, and an extrinsic branching criterion is not needed [30,31].

Due to the highly heterogeneity of the microstructural, mechanical, and fracture properties, welded joints are a vulnerable component of structures, and they are prone to pores, cracks, and other defects. However, compared with the homogeneous materials that are mainly studied and mentioned above, the crack-branching behavior was rarely reported in welded joints. Understanding the branching criterion or mechanism governing the bifurcation of a crack in welded joints is still a challenge.

In the fracture mechanics experiment for a dissimilar metal welded joint (DMWJ), which is used for connecting the pipe nozzle and the safe end in nuclear power plants, the authors found that the crack-branching behavior occurred occasionally [32,33]. However, it is not clear when the crack will surely branch in welded joints, nor the crack-branching mechanism. Thus, in this research, three kinds of crack-branching models that reflect simplified welded joints were designed; the aim of the present paper works on finding and capturing the crack-branching behavior in welded joints and shedding light on its branching mechanism.

2. The Designed Crack-Branching Model

Since the crack deviation phenomenon occurs in the crack propagation process under local strength mismatch, and the crack deviates to the side of material with lower strength [32,33], it can be assumed that the crack will branch when there are similar strength mismatches on two sides of the crack. Based on this assumption, three kinds of simple crack-branching models, which can reflect simplified welded joints, were designed to capture the crack-branching behavior, as presented in Sections 2.1–2.3.

All these models were composed by three regions: the left region, the center region, and the right region. Different material properties, which are obtained by changing the true stress versus strain curve of a ductile material, ferrite low-alloy steel A508, were assigned to the three regions. Figure 1 presents the true stress versus strain curve of the ductile material A508 [34]. Its elastic modulus E is 202,410 MPa, and Poisson's ratio v is 0.3.

For all models, a load roll is applied at the top and center of the model, and two back-up rolls are applied at the bottom of the model. The loading is applied at the load roll by prescribing a displacement of 30 mm, and the two back-up rolls are fixed by control displacement and rotation. The initial crack is located in the middle of the model. The model width is 32 mm (W = 32 mm), the loading span is 128 mm (L = 4W), and the initial crack length is 16 mm (a/W = 0.5). All models are two-dimensional (2D) plane strain specimen models. Compared with the three-dimensional (3D) specimen models, the 2D plane strain specimen with W = 32 mm, L = 4W, and a/W = 0.5 has the same J-resistance curve with the 3D specimen with W = 32 mm, L = 4W, a/W = 0.5, and B/W = 0.5.

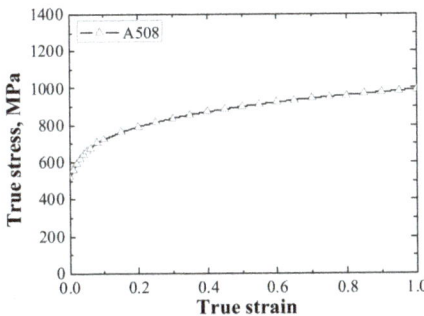

Figure 1. The true stress versus strain curve of the ductile material A508.

2.1. The First Crack-Branching Model

In the first crack-branching model, the material of the center region is fixed as the A508, and its true stress versus strain curve was assigned to this region. The left region and the right region have the same material properties, and 0.2–2 times (0.2, 0.4, 0.6, 0.8, 1.0, 1.2, 1.4, 1.6, 1.8, 2.0) of the true stress versus strain curves of the ductile material A508 were assigned to them successively, as shown in Figure 2. In addition, the widths of the left, center, and right regions are fixed to 64, 20, and 64 mm, respectively.

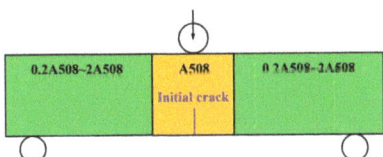

Figure 2. The structure and materials of the first crack-branching model.

2.2. The Second Crack-Branching Model

In the second crack-branching model, both materials of the center region and the right region are fixed, and one time and two times of the true stress versus strain curves of the ductile material A508 were assigned to them, respectively. In addition, 0.2–1 times (0.2, 0.3, 0.4, 0.5, 0.6, 0.7, 0.8, 0.9, 1.0) of the true stress versus strain curves of the ductile material A508 were assigned to the left region successively, as shown in the Figure 3. In addition, the widths of left, center, and right regions are fixed to 64, 20, and 64 mm, respectively.

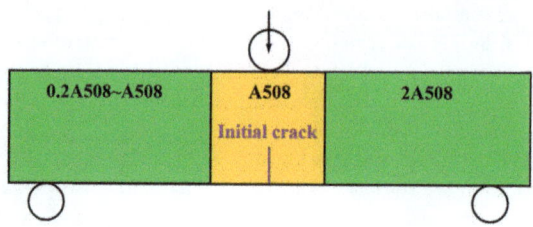

Figure 3. The structure and materials of the second crack-branching model.

2.3. The Third Crack-Branching Model

In the third crack-branching model, all the materials of the left region, the center region, and the right region are fixed. Twice the true stress versus strain curve of the ductile material A508 was assigned to the center region, and half of the true stress versus strain curve of the ductile material A508 was assigned to the left region and the right region. Different from the first and second models whose widths of different regions are fixed, the center region width in the third model was changing from 10 to 90 mm (10, 20, 30, 40, 50, 60, 70, 80, and 90 mm), as shown in Figure 4.

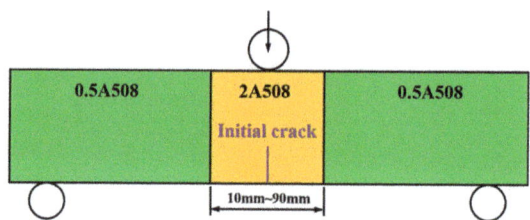

Figure 4. The structure and materials of the third crack-branching model.

3. Finite Element Numerical Calculation

Crack propagation in ductile metals is a complex multiscale phenomenon governed by the initiation, growth, and coalescence of micro-voids. To observe the crack-branching behavior, the finite element analysis based on the Gurson–Tvergaard–Needleman (GTN) damage model [35–39] was chosen to obtain and monitor the whole crack propagation process in different designed crack-branching models.

The yield function of the GTN damage model was expressed as

$$\phi(\sigma_m, \sigma_{eq}, f^*) = \frac{\sigma_{eq}^2}{\sigma_f^2} + 2q_1 f^* \cosh\left(\frac{3q_2 \sigma_m}{2\sigma_f}\right) - 1 - q_3 f^{*2} = 0 \qquad (1)$$

where q_1, q_2, and q_3 are parameters determined by ad hoc finite element (FE) simulations, in which σ_m is the mean stress, σ_{eq} is the equivalent stress, σ_f is the flow stress. The f^* is the void volume fraction (VVF), and it is the replacement of f in the Gurson model. The relationship of them was expressed as

$$f^* = \begin{cases} f & \text{if } f \leq f_c \\ f_c + \frac{f_F^* - f_c}{f_F - f_c}(f - f_c) & \text{if } f_c < f < f_F \\ f_F^* & \text{if } f \geq f_F \end{cases} \qquad (2)$$

where f_C is the critical VVF, f_F is the final failure parameter, and f_F^* is calculated as

$$f_F^* = \left(q_1 + \sqrt{q_1^2 - q_3}\right)/q_3. \qquad (3)$$

The change in VVF during an increment of deformation contains two parts: one due to the growth of existing voids, and the other due to the nucleation of new voids.

$$\dot{f} = \dot{f}_{growth} + \dot{f}_{nucleation} \qquad (4)$$

where

$$\dot{f}_{growth} = (1-f)\dot{\varepsilon}^P_{kk} \qquad (5)$$

and

$$\dot{f}_{nucleation} = A\dot{\varepsilon}^p = \frac{f_N}{S_N \sqrt{2\pi}} \exp[-\frac{1}{2}(\frac{\varepsilon^p - \varepsilon_N}{S_N})^2]\dot{\varepsilon}^p. \qquad (6)$$

here, $\dot{f}$ is the VVF growth rate, $\dot{f}_{growth}$ is the VVF growth rate due to the growth of existing voids, $\dot{f}_{nucleation}$ is the VVF growth rate due to the nucleation of new voids, $\dot{\varepsilon}^P_{kk}$ is the change rate of plastic strain, ε^p is equivalent plastic strain, and $\dot{\varepsilon}^p$ is the change rate of equivalent plastic strain.

Generally, it contains nine parameters in the GTN damage model: q_1, q_2, q_3, ε_N, S_N, f_N, f_0, f_C, and f_F. Of these, q_1, q_2, and q_3 are parameters determined by ad hoc FE simulations; ε_N, S_N, and f_N are void nucleation parameters; f_0 is the initial VVF; f_C is the critical VVF; and f_F is the final failure parameter. When the VVF reaches the critical value f_C, void coalescence occurs. When the VVF reaches the final value f_F, fracture occurs. For the material A508, $q_1 = 1.5$, $q_2 = 1$, $q_3 = 2.25$, ε_N =0.3, $S_N = 0.1$, $f_N = 0.002$, $f_0 = 0.0002$, $f_C = 0.04$, and $f_F = 0.17$ [40,41].

The GTN damage model has been implemented in the ABAQUS code (6.14, Dassault Systèmes group company, Shanghai, China), and it was widely selected to obtain the crack propagation process [42–45]. Figure 5 presents the finite element meshes of the typical designed crack-branching model. The 2D plane strain four-node isoperimetric elements with reduced integration (CPE4R) was used, and the mesh size in the crack propagation region is 0.05 mm × 0.1 mm [42,43]. The crack propagation path in the propagation process can be observed from the finite element method simulation results directly.

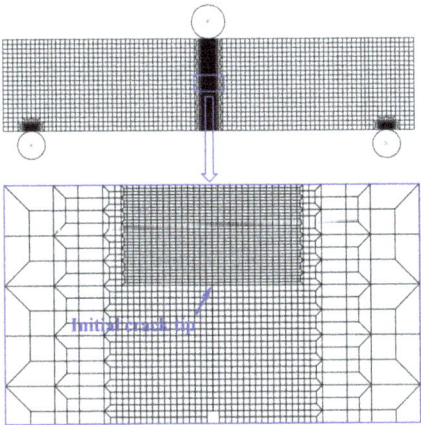

Figure 5. The finite element meshes of the typical designed crack-branching model.

4. Verification of the Gurson-Tvergaard-Needleman Damage Model by Experimental Results

To ensure the accuracy of the finite element results obtained by the GTN damage model, some experiments have been performed and compared with the finite element analysis in the previous studies [44,46]. In the experiments, an Alloy52M dissimilar metal welded joint (DMWJ) that contains

A508, austenitic stainless steel 316L, buttering layer material Alloy52Mb, and weld metal material Alloy52Mw was selected, and the single edge notched bend (SENB) specimens with five crack depths denoted as a/W = 0.2, 0.3, 0.5, 0.6, and 0.7 were manufactured from the DMWJ. The crack propagation paths and J-resistance curves of different SENB specimens were obtained and compared with the results obtained by the GTN damage model. The experiments were carried out by an Instron screw-driven machine at room temperature. The quasi-static loading was conducted by displacement controlled mode at a cross-head speed of 0.5 mm/min, and the load–load line displacement curves were automatically recorded by a computer aided control system of the testing machine. Figure 6 presents one of the comparisons of experimental results with finite element results for the SENB specimen with a/W = 0.5 [44,46]. It clearly demonstrates that the finite element result obtained by the GTN damage model is accurate, and the GTN damage model can be used to simulate the crack propagation process and obtain the crack propagation path.

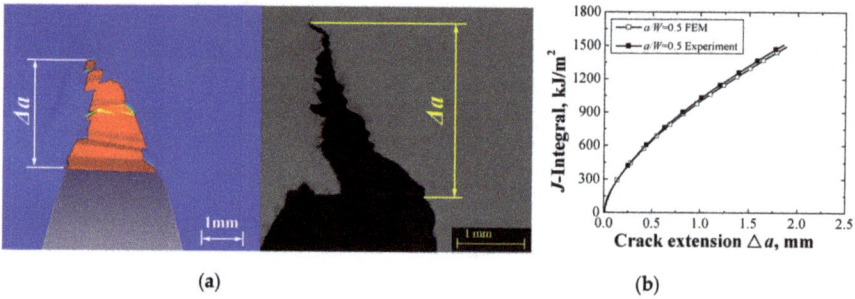

Figure 6. The comparison of finite element results obtained by the Gurson–Tvergaard–Needleman (GTN) damage model with experimental results: (**a**) crack propagation paths; (**b**) J-resistance curves.

5. Results and Discussion

5.1. The First Crack-Branching Model

Figure 7 presents the crack propagation paths of all the first crack-branching models. Figure 8 presents the typical crack propagation paths when the models with 0.6 and 0.8 times the true stress versus strain curves of the ductile material A508 were assigned to the left region and the right region.

It can be found from Figures 7 and 8 that when the model with 0.2 times the true stress versus strain curve of the ductile material A508 was assigned to the left region and the right region, the single crack splits into four branches under the highest strength mismatch. One of the branches deviates a little bit to the left side, one of the branches deviates a little bit to the right side, and the other two branches deviate to the left and right sides separately and grow along with the directions perpendicular to the initial crack. When the models with 0.4 and 0.6 times the true stress versus strain curves of the ductile material A508 were assigned to the left region and the right region, the single crack splits into three branches, as shown in Figure 8a. With decreasing strength mismatch, the branching phenomenon becomes weak. When the model with 0.8 times the true stress versus strain curve of the ductile material A508 was assigned to the left region and the right region, the single crack splits into two branches. One of the branches deviates a little bit to the left side, and one of the branches deviates a little bit to the right side, as shown in Figure 8b. Furthermore, when the model with equal to or higher than one time the true stress versus strain curve of the ductile material A508 was assigned to the left region and right region, the crack does not branch.

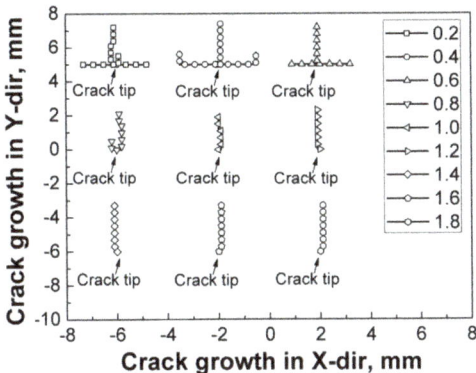

Figure 7. The crack propagation paths of all the first crack-branching models.

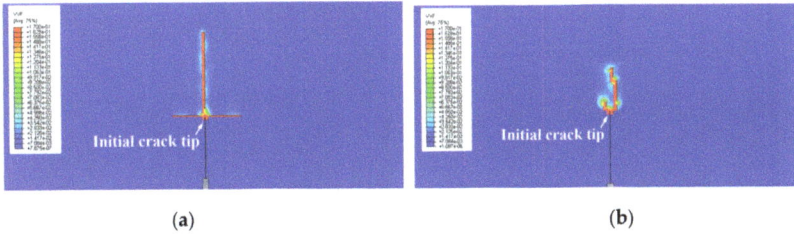

Figure 8. The typical crack propagation paths when 0.6 times (**a**) and 0.8 times (**b**) the true stress versus strain curves of the ductile material A508 were assigned to the left region and the right region.

The results presented clearly demonstrate that the high strength mismatch induces branching. Since the crack growth path deflects to the side with low strength, when the strength mismatches on the left and right sides of the crack are similar, the crack has a tendency to spread to both the left and right sides and certainly in the direction of the initial crack. Under the combined action of them, the crack-branching behavior occurs. In addition, with gradual increasing of the strength mismatch, the crack-branching behavior becomes more apparent. It is not just the quantity of the branches that increases: the branch can even grow along with the direction perpendicular to the initial crack. When the material strengths of the left and right regions are equal to or higher than the center region where the initial crack is located, the crack-branching behavior does not occur.

5.2. The Second Crack-Branching Model

Figure 9 presents the crack propagation paths of all the second crack-branching models. Figure 10 presents the typical crack propagation paths when the models with 0.6 and 0.8 times the true stress versus strain curves of the ductile material A508 were assigned to the left region.

Different from the first crack-branching models, the material properties in the second crack-branching models are changing gradually from left to right. The left region has the lowest strength, while the right region has the highest strength. It can be found from Figures 9 and 10 that when the models with 0.2–0.7 times the true stress versus strain curves of the ductile material A508 were assigned to the left region, the single crack splits into two branches. One of the branches deviates to the left side and grows along the direction perpendicular to the initial crack under high strength mismatch, and the propagation direction of the other branch changes from deviating a little bit to the left side to along the direction of the initial crack, as shown in Figure 10a. When the model with 0.8 times the true stress versus strain curve of the ductile material A508 was assigned to the left region, although it still has two crack branches, the branching phenomenon becomes weak under low strength mismatch. One of the short branches grows along the direction of the initial crack, and the other long branch deviates a little bit to the left side, as shown in Figure 10b. When the models with 0.9–1 times the true stress versus strain curves of the ductile material A508 were assigned to the left region, the crack does not branch.

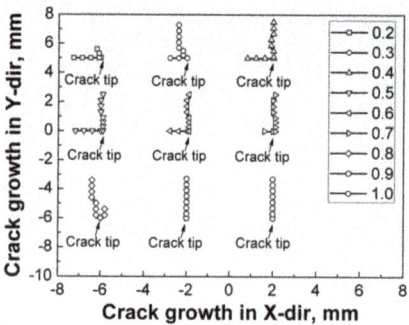

Figure 9. The crack propagation paths of all the second crack-branching models.

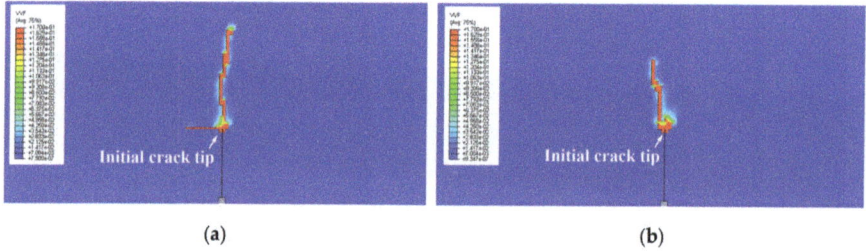

Figure 10. The typical crack propagation paths when the models with 0.6 times (**a**) and 0.8 times (**b**) the true stress versus strain curves of the ductile material A508 were assigned to the left region.

Since the strength mismatches on both the left and right sides of the crack are different in this kind of model, the results demonstrate that the similar strength mismatch on each side of the crack, which is mentioned in Section 5.1, is a sufficient but not necessary condition for crack branching. As long as there is another large enough propagation trend that is different from the original crack propagation direction, the crack-branching behavior occurs. With increasing strength mismatch, the branching trend becomes more apparent.

5.3. The Third Crack-Branching Model

Figure 11 presents the crack propagation paths of all the third crack-branching models. The typical crack propagation paths when the widths of the center region are 20 and 50 mm are shown in Figure 12, respectively.

Different from the first and second crack-branching models, Figures 11 and 12 present the crack-branching behavior under different center region widths. It found that when the center region width is 10 mm, the single crack splits into two branches. Since the center region is too narrow, the two branches do not propagate along the initial crack direction, but they deviate to the left and right sides separately and grow along with the direction perpendicular to the initial crack. When the center region width is 20 mm, the single crack splits into two branches firstly; then, more crack-branching behaviors occur in the crack propagation process. When the center region width is longer than 20 mm, the single crack splits into two branches. One of the branches deviates a little bit to the left side, and the other branch deviates a little bit to the right side. Especially, when the center region width is longer than

70 mm, the crack branch behaviors do not change. This is because there exists an effect range of material constraint in the welded joints. When the center region width is longer than 70 mm, it exceeds this effect range, and the fracture behaviors of weld joints are no longer influenced by the material mechanical properties that are located out of this effect range [47–49].

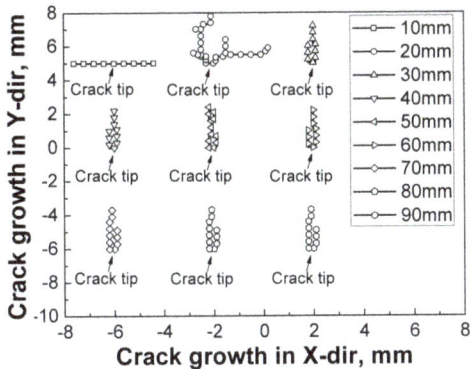

Figure 11. The crack propagation paths of all the third crack-branching models.

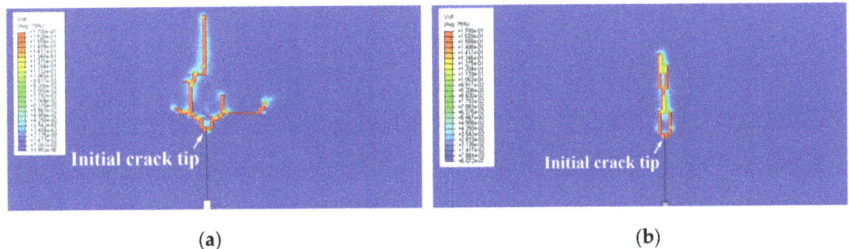

Figure 12. The typical crack propagation paths when the widths of the center region were 20 mm (**a**) and 50 mm (**b**).

The results presented also demonstrate that the crack-branching behavior is also affected by the dimensions of different regions in welded joints. A high strength mismatch that is induced by both mechanical properties and dimensions of different regions is the key of crack branching in welded joints.

5.4. Crack-Branching Mechanism

In all the above models, the third crack-branching model has the most complex crack-branching behavior when the center region width is 20 mm, as shown in Figure 12a. Thus, in this section, this model was selected to analyze the crack-branching mechanism.

Figure 13 presents the VVF and stress distributions at the time of crack branching. Figure 13a presents the status when the first crack-branching event occurs, and Figure 13b presents the status when the second and third crack-branching events occur. It is found that each crack branching is accompanied by three local high stress concentrations at crack tip, which are produced by loading and strength mismatch. Three pulling forces are created by the three local high stress concentrations that pull the crack propagation along with the directions of stress concentrations.

It can also be found that one of the pulling forces lies in the original crack direction; the other two pulling forces lie in the two sides of the original crack, and they have a similar angle to that of the original crack direction. Under the combined action of the three pulling forces, crack branching occurs, and two new cracks are initiated from the middle of the pulling forces.

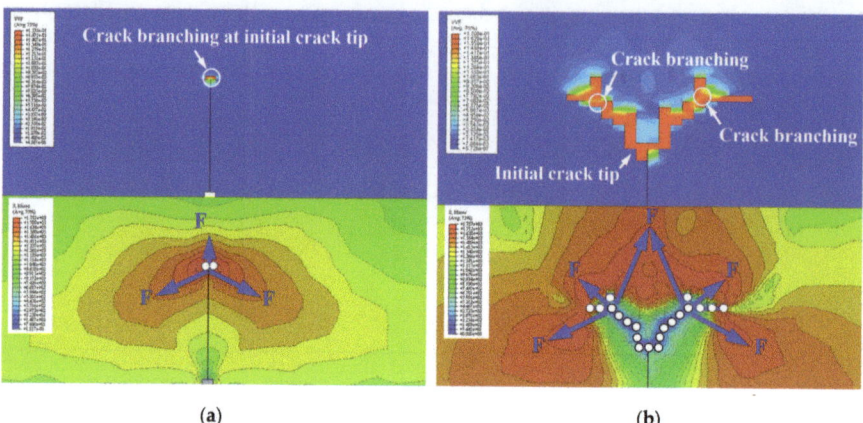

Figure 13. The void volume fraction (VVF) and stress distributions when the first crack-branching event occurs (**a**) and the second and third crack-branching events (**b**).

In general, the crack-branching behavior (Supplementary Materials) in welded joints was captured and analyzed in this study. Compared with the results obtained from the three kinds of crack-branching models, it can be found that the crack-branching behavior is closely related to high strength mismatch in welded joints. The high strength mismatch may be caused by the mechanical properties of the different regions (the first crack-branching model and the second crack-branching model), or it may be caused by the dimensions of the different regions (the third crack-branching model). With increasing strength mismatch, a crack may split into two, three, or more branches. Moreover, the branch can even grow along with the direction perpendicular to the initial crack under higher strength mismatch. These phenomena are dangerous for welded joints and need to be paid close attention.

In addition, the crack-branching mechanism was also analyzed in this study. Each crack branching is accompanied by three local high stress concentrations at the crack tip. The three local high stress concentrations were produced by loading and strength mismatch, and they will create three pulling forces. The pulling forces will give different propagation trends to the crack. Under the combined action of the three pulling forces, crack branching occurs.

However, there are also some deficiencies. The experimental results that correspond to the crack-branching models were not contained in this study. The main reason is that there exists a heat-affected zone, fusion zone, and near-interface zone in the welded joints. It is difficult to manufacture ideal welded joints the same as the designed models. Thus, for elaborating the problem simply and clearly, some basic models were designed, and only the finite element method was used in this study to obtain the crack-branching behavior. It is a pity that the results in this study do not provide an exact mechanical justification and crack-branching criterion.

In the future, the crack-branching criterion and the correlation of the crack-branching behavior with strength mismatch should be established, and the crack-branching behavior of welded joints can be judged and obtained directly rather than by finite element method simulations or experiments. Then, the results can provide scientific support for the structural integrity assessment and the design of welded joints.

6. Conclusions

(1) With gradual increasing of the strength mismatch, crack-branching behavior becomes more apparent. Not just the quantity of the branches increases, the branch can even grow along with the direction perpendicular to the initial crack.

(2) The similar strength mismatch on each side of the crack is a sufficient but not necessary condition for crack branching. As long as there is another large enough propagation trend that is different from the original crack propagation direction, crack-branching behavior occurs. A high strength mismatch that is induced by both the mechanical properties and dimensions of different regions is the key of crack branching in welded joints.

(3) Each crack branching is accompanied by three local high stress concentrations at the crack tip. Three pulling forces that are created by the three local high stress concentrations pull the crack propagation along with the directions of stress concentrations. Under the combined action of the three pulling forces, crack branching occurs, and two new cracks initiate from the middle of the pulling forces.

(4) The finite element method based on the GTN damage model is an effective method to simulate the crack-branching behavior in welded joints during the crack propagation process.

Supplementary Materials: The following are available online at http://www.mdpi.com/2075-4701/10/10/1308/s1, Video S1: crack branching behavior.

Author Contributions: Conceptualization, J.Y.; methodology, J.Y.; software, W.W. and Q.Y.; validation, W.W., J.Y., H.C. and Q.Y.; writing—original draft preparation, W.W.; writing—review and editing, J.Y. and H.C. All authors have read and agreed to the published version of the manuscript.

Funding: This research was funded by (National Natural Science Foundation of China) grant number (51975378 and 51828501).

Acknowledgments: The authors would like to thank University of Shanghai for Science and Technology, University of Strathclyde and Shanghai Municipal Education Commission for their support.

Conflicts of Interest: The authors declare no conflict of interest.

References

1. Cox, B.N.; Gao, H.; Gross, D.; Rittel, D. Modern topics and challenges in dynamic fracture. *J. Mech. Phys. Solids* **2005**, *53*, 565–596. [CrossRef]
2. Bouchbinder, E.; Fineberg, J.; Marder, M. Dynamics of simple cracks. *Annu. Rev. Condens. Matter Phys.* **2010**, *1*, 371–395. [CrossRef]
3. Forquin, P. An optical correlation technique for characterizing the crack velocity in concrete. *Eur. Phys. J. Spec. Top.* **2012**, *206*, 89–95. [CrossRef]
4. Curbach, M.; Eibl, J. Crack velocity in concrete. *Eng. Fract. Mech.* **1990**, *35*, 321–326. [CrossRef]
5. Ožbolt, J.; Sharma, A.; Reinhardt, H.W. Dynamic fracture of concrete-compact tension specimen. *Int. J. Solids Struct.* **2011**, *48*, 1534–1543. [CrossRef]
6. Ožbolt, J.; Bošnjak, J.; Sola, E. Dynamic fracture of concrete compact tension specimen: Experimental and numerical study. *Int. J. Solids Struct.* **2013**, *50*, 4270–4278. [CrossRef]
7. Ožbolt, J.; Bede, N.; Sharma, A.; Mayer, U. Dynamic fracture of concrete L-specimen: Experimental and numerical study. *Eng. Fract. Mech.* **2015**, *148*, 27–41. [CrossRef]
8. Zhang, Q.B.; Zhao, J. A review of dynamic experimental techniques and mechanical behaviour of rock materials. *Rock Mech.* **2014**, *47*, 1411–1478. [CrossRef]
9. Mecholsky, J.J., Jr.; DeLellis, D.P.; Mecholsky, N.A. Relationship between fractography, fractal analysis and crack branching. *J. Eur. Ceram. Soc.* **2020**, *40*, 4722–4726. [CrossRef]
10. Nakamura, N.; Kawabata, T.; Takashima, Y.; Yanagimoto, F. Effect of the stress field on crack branching in brittle material. *Theor. Appl. Fract. Mech.* **2020**, *128*, 102583. [CrossRef]
11. Chen, Z.; Ju, J.W.; Su, G.; Huang, X.; Li, S.; Zhai, L. Influence of micro-modulus functions on peridynamics simulation of crack propagation and branching in brittle materials. *Eng. Fract. Mech.* **2019**, *216*, 106498. [CrossRef]
12. Kou, M.M.; Lian, Y.J.; Wang, Y.T. Numerical investigations on crack propagation and crack branching in brittle solids under dynamic loading using bond-particle model. *Eng. Fract. Mech.* **2019**, *212*, 41–56. [CrossRef]
13. Li, J.; Xie, Y.J.; Zheng, X.Y.; Cai, Y.M. Underlying fracture trends and triggering on Mode-II crack branching and kinking for quasi-brittle solids. *Eng. Fract. Mech.* **2019**, *211*, 382–400. [CrossRef]

14. Bouchbinder, E.; Mathiesen, J.; Procaccia, I. Branching instabilities in rapid fracture: Dynamics and geometry. *Phys. Rev. E* **2005**, *71*, 056118. [CrossRef] [PubMed]
15. Boué, T.G.; Cohen, G.; Fineberg, J. Origin of the microbranching instability in rapid cracks. *Phys. Rev. Lett.* **2015**, *114*, 054301. [CrossRef]
16. Karma, A.; Lobkovsky, A.E. Unsteady crack motion and branching in a phase-field model of brittle fracture. *Phys. Rev. Lett.* **2004**, *92*, 245510. [CrossRef] [PubMed]
17. Falk, M.L.; Needleman, A.; Rice, J.R. A critical evaluation of cohesive region models of dynamic fracture. *J. Phys. IV* **2001**, *11*, 43–50.
18. Zhou, F.; Molinari, J.F.; Shioya, T. A rate-dependent cohesive model for simulating dynamic crack propagation in brittle materials. *Eng. Fract. Mech.* **2005**, *72*, 1383–1410. [CrossRef]
19. Belytschko, T.; Chen, H.; Xu, J.; Zi, G. Dynamic crack propagation based on loss of hyperbolicity and a new discontinuous enrichment. *Int. J. Numer. Meth. Eng.* **2003**, *58*, 1873–1905. [CrossRef]
20. Xu, D.; Liu, Z.; Liu, X.; Zeng, Q.; Zhuang, Z. Modeling of dynamic crack branching by enhanced extended finite element method. *Comput. Mech.* **2014**, *54*, 489–502. [CrossRef]
21. Henry, H. Study of the branching instability using a phase field model of inplane crack propagation. *EPL-Europhys. Lett.* **2008**, *83*, 16004. [CrossRef]
22. Bleyer, J.; Roux-Langlois, C.; Molinari, J.F. Dynamic crack propagation with a variational phase-field model: Limiting speed, crack branching and velocity-toughening mechanisms. *Int. J. Fract.* **2017**, *204*, 79–100. [CrossRef]
23. Hofacker, M.; Miehe, C. Continuum phase field modeling of dynamic fracture: Variational principles and staggered FE implementation. *Int. J. Fract.* **2012**, *178*, 113–129. [CrossRef]
24. Hofacker, M.; Miehe, C. A phase field model of dynamic fracture: Robust field updates for the analysis of complex crack patterns. *Int. J. Numer. Meth. Eng.* **2013**, *93*, 276–301. [CrossRef]
25. Karma, A.; Kessler, D.A.; Levine, H. Phase-field model of mode III dynamic fracture. *Phys. Rev. Lett.* **2001**, *87*, 045501. [CrossRef] [PubMed]
26. Henry, H.; Adda-Bedia, M. Fractographic aspects of crack branching instability using a phase-field model. *Phys. Rev. E* **2013**, *88*, 060401. [CrossRef]
27. Bobaru, F.; Zhang, G. Why do cracks branch? A peridynamic investigation of dynamic brittle fracture. *Int. J. Fract.* **2015**, *196*, 59–98. [CrossRef]
28. Lloberas-Valls, O.; Huespe, A.E.; Oliver, J.; Dias, I.F. Strain injection techniques in dynamic fracture modeling. *Comput. Method Appl. Mech.* **2016**, *308*, 499–534. [CrossRef]
29. Linder, C.; Armero, F. Finite elements with embedded branching. *Finite Elem. Anal. Des.* **2009**, *45*, 280–293. [CrossRef]
30. Geelen, R.J.M.; Liu, Y.; Dolbow, J.E.; Rodríguez-Ferran, A. An optimization-based phase-field method for continuous-discontinuous crack propagation. *Int. J. Numer. Meth. Eng.* **2018**, *116*, 1–20. [CrossRef]
31. Wu, T.; Carpiuc-Prisacari, A.; Poncelet, M.; De Lorenzis, L. Phase-field simulation of interactive mixed-mode fracture tests on cement mortar with full-field displacement boundary conditions. *Eng. Fract. Mech.* **2019**, *182*, 658–688. [CrossRef]
32. Yang, J.; Wang, G.Z.; Xuan, F.Z.; Tu, S.T.; Liu, C.J. An experimental investigation of in-plane constraint effect on local fracture resistance of a dissimilar metal welded joint. *Mater. Des.* **2014**, *53*, 611–619. [CrossRef]
33. Yang, J.; Wang, G.Z.; Xuan, F.Z.; Tu, S.T.; Liu, C.J. Out-of-plane constraint effect on local fracture resistance of a dissimilar metal welded joint. *Mater. Des.* **2014**, *55*, 542–550. [CrossRef]
34. Wang, H.T.; Wang, G.Z.; Xuan, F.Z.; Liu, C.J.; Tu, S.T. Local mechanical properties and microstructures of Alloy52M dissimilar metal welded joint between A508 ferritic steel and 316L stainless steel. *Adv. Mater. Res.* **2012**, *509*, 103–110. [CrossRef]
35. Gurson, A.L. Continuum theory of ductile rupture by void nucleation and growth: Part I-Yield criteria and flow rules for porous ductile media. *J. Eng. Mater. Technol.* **1977**, *99*, 2–15. [CrossRef]
36. Chu, C.C.; Needleman, A. Void nucleation effects in biaxially stretched sheets. *J. Eng. Mater. Technol.* **1980**, *102*, 249–256. [CrossRef]
37. Tvergaard, V. Influence of voids on shear band instabilities under plane strain conditions. *Int. J. Fract.* **1981**, *17*, 389–407. [CrossRef]
38. Tvergaard, V. On localization in ductile materials containing spherical voids. *Int. J. Fract.* **1982**, *18*, 157–169.

39. Tvergaard, V.; Needleman, A. Analysis of the cup-cone fracture in a round tensile bar. *Acta Metal.* **1984**, *32*, 157–169. [CrossRef]
40. Benseddiq, N.; Imad, A. A ductile fracture analysis using a local damage model. *Int. J. Pres. Ves. Pip.* **2008**, *85*, 219–227. [CrossRef]
41. Yang, J.; Wang, G.Z.; Xuan, F.Z.; Tu, S.T. Unified characterisation of in-plane and out-of-plane constraint based on crack-tip equivalent plastic strain. *Fatigue Fract. Eng. Mater. Struct.* **2013**, *36*, 504–514. [CrossRef]
42. Østby, E.; Thaulow, C.; Zhang, Z.L. Numerical simulations of specimen size and mismatch effects in ductile crack growth-Part I: Tearing resistance and crack growth paths. *Eng. Fract. Mech.* **2007**, *74*, 1770–1792. [CrossRef]
43. Østby, E.; Thaulow, C.; Zhang, Z.L. Numerical simulations of specimen size and mismatch effects in ductile crack growth-Part II: Near-tip stress fields. *Eng. Fract. Mech.* **2007**, *74*, 1793–1809. [CrossRef]
44. Yang, J. Micromechanical analysis of in-plane constraint effect on local fracture behavior of cracks in the weakest locations of dissimilar metal welded joint. *Acta Metall. Sin.* **2017**, *30*, 840–850. [CrossRef]
45. Penuelas, I.; Betegon, C.; Rodriguez, C. A ductile failure model applied to the determination of the fracture toughness of welded joints. Numerical simulation and experimental validation. *Eng. Fract. Mech.* **2006**, *73*, 2756–2773. [CrossRef]
46. Wang, H.T.; Wang, G.Z.; Xuan, F.Z.; Tu, S.T. An experimental investigation of local fracture resistance and crack growth paths in a dissimilar metal welded joint. *Mater. Des.* **2013**, *44*, 179–189. [CrossRef]
47. Yang, J.; Wang, L. Effect Range of the Material Constraint—I. Center Crack. *Materials* **2019**, *12*, 67. [CrossRef]
48. Dai, Y.; Yang, J.; Wang, L. Effect Range of the Material Constraint—II. Interface Crack. *Metals* **2019**, *9*, 696. [CrossRef]
49. Dai, Y.; Yang, J.; Chen, H.F. Effect range of the material constraint in different strength mismatched laboratory specimens. *Appl. Sci.* **2020**, *10*, 2434. [CrossRef]

© 2020 by the authors. Licensee MDPI, Basel, Switzerland. This article is an open access article distributed under the terms and conditions of the Creative Commons Attribution (CC BY) license (http://creativecommons.org/licenses/by/4.0/).

Article

An Ultra-High Frequency Vibration-Based Fatigue Test and Its Comparative Study of a Titanium Alloy in the VHCF Regime

Wei Xu [1,*], Yanguang Zhao [2], Xin Chen [1], Bin Zhong [1], Huichen Yu [1], Yuhuai He [1] and Chunhu Tao [1]

[1] Beijing Key Laboratory of Aeronautical Materials Testing and Evaluation, AECC Key Laboratory of Science Technology on Aeronautical Materials Testing and Evaluation, Science and Technology on Advanced High Temperature Structural Materials Laboratory, Beijing Institute of Aeronautical Materials, Beijing 100095, China; sxdhchenxin@foxmail.com (X.C.); zhongbin@biam.ac.cn (B.Z.); yuhuichen@biam.ac.cn (H.Y.); heyuhuai@biam.ac.cn (Y.H.); taochunhu@biam.ac.cn (C.T.)

[2] State Key Laboratory of Structural Analysis of Industrial Equipment, Dalian University of Technology, Dalian 116085, China; ygzhao81@dlut.edu.cn

* Correspondence: wxu621@163.com; Tel.: +86-10-62496728

Received: 1 October 2020; Accepted: 22 October 2020; Published: 24 October 2020

Abstract: This paper proposes an ultra-high frequency (UHF) fatigue test of a titanium alloy TA11 based on electrodynamic shaker in order to develop a feasible testing method in the VHCF regime. Firstly, a type of UHF fatigue specimen is designed to make its actual testing frequency reach as high as 1756 Hz. Then the influences of the loading frequency and loading types on the testing results are considered separately, and a series of comparative fatigue tests are hence conducted. The results show the testing data from the present UHF fatigue specimen agree well with those from the conventional vibration fatigue specimen with the loading frequency of 240 Hz. Furthermore, the present UHF testing data show good consistency with those from the axial-loading fatigue and rotating bending fatigue tests. But the obtained fatigue life from ultrasonic fatigue test with the loading frequency of 20 kHz is significantly higher than all other fatigue test results. Thus the proposed ultra-high frequency vibration-based fatigue test shows a balance of high efficiency and similarity with the conventional testing results.

Keywords: vibration-based fatigue; ultra-high frequency; very high cycle fatigue; fatigue test; titanium alloy

1. Introduction

Aviation equipments, such as aeroplanes and aeroengines always undergo cyclic stress during service time, thus fatigue damage has been a major concern in the researches of aeronautical materials and structures. In recent decades, the academic and engineering communities have gradually realized that fatigue fracture of materials can occur after 10^7 cycles or even 10^8 cycles. Especially for aviation equipment, the failure forms of many structural components belong to very high cycle fatigue (VHCF) regime. As a result, VHCF has gradually been paid an increasing number of attentions in the design of the aviation equipments [1], with the higher requirements of service life and reliability.

Fatigue testing is an essential aspect in fatigue researches. Due to the ultra-high failure cycles, improving the loading efficiency is very crucial in the VHCF testing. Several testing equipments have been used for the testing of VHCF, such as rotating-bending fatigue tester, servo-hydraulic fatigue tester, electromagnetic resonance tester and ultrasonic fatigue tester. The first three types of the testing approaches are usually regarded as conventional fatigue testers, which have the general loading frequency range within 10 to 100 Hz. Thus it would be extremely time-consuming using the conventional fatigue testers in the VHCF regime [2]. In contrast, the development of the ultrasonic fatigue tester has greatly improved the loading efficiency of the VHCF testing since the actual loading frequency reaches up to 20 kHz [3]. Benefit from the huge progress in fatigue testing efficiency, the researches in VHCF of titanium alloys and other metals have been widely conducted and some interesting fatigue testing data and failure mechanism have been obtained [4–6].

However, the significant increase in loading frequency would remarkably influence the fatigue strength and life of some materials in the VHCF regime, which have been reported by many researchers [2,4,7] Also it is controversial whether the fatigue failure mechanism under the conditions of ultrasonic fatigue testing is similar to those of conventional high-frequency fatigue testing [8,9]. Meanwhile, some investigations have supported that the results from the ultrasonic fatigue tester were close to those from the conventional fatigue testers [2]. Anyway, it can be stated that the application of the ultrasonic fatigue testing in the VHCF regime is still in controversy. Furthermore, a widely-accepted testing standard of the ultrasonic fatigue has not been proposed, while the currently existing testing standards (e.g., ISO and ASTM standards) are only applicable for the conventional fatigue tests.

For aviation engineering, VHCF issues are generally introduced by the vibration of the moving components such as blades and vanes. When subjected to fatigue loadings during the working condition, these components would always experience bending or twist loads at high frequencies [10,11]. For one thing, the actual working frequency of these aviation components cannot reach as high as the loading frequency (i.e., 20 kHz) of the ultrasonic fatigue testing. For another thing, axial-loading fatigue data does not provide a sufficient representation of HCF or VHCF behaviors of the vibrating components [12]. These situations are not adequately simulated by axial-loading fatigue tests. In short, both of the conventional axial loading fatigue testing and the ultrasonic fatigue testing data are insufficient for the design of the vibrating components in the VHCF regime.

Accordingly, vibration-based fatigue tests have been developed and carried out to obtain more meaningful data for the vibrating components. Also it is a proper testing approach to study the bending fatigue properties within the reduced experimental period since the testing frequency is much larger than the conventional axial-loading fatigue tests [13]. In other words, the vibration-based fatigue test is a sort of speeding-up fatigue test. Vibration-based bending fatigue tests have been usually carried out by electrodynamic shakers. It is widely known that the response amplitude of a specimen reaches its maximum value when the specimen vibrates under resonance condition for same excitation amplitude [14]. Thus the specimens are usually excited in a high frequency resonant mode for the purpose of reducing the power-consuming of the testing system [15,16]. This can be supported by several vibration fatigue studies of different materials [17,18], most of which have been carried out in the regime of conventional fatigue cycle though. And few work of vibration fatigue has been reported in the VHCF regime of materials.

This research proposes an experimental method for ultra-high frequency fatigue of materials using an electrodynamic shaker. For a type of titanium alloy commonly used in aero engines, an ultra-high frequency (UHF) fatigue specimen is independently designed and the fatigue experiments in the HCF and VHCF regimes are conducted. Furthermore, the influences of the loading frequency and loading types on the testing results of the fatigue life are considered separately, and a series of comparative fatigue tests are hence conducted, with the testing results compared with the present UHF results finally.

2. UHF Fatigue Specimen Design

The fatigue testing method used in the present study is actually based on resonance. Thus the loading frequency of the present fatigue testing system is very close to the resonance frequency of the specimen. It is widely known that the number of the vibration mode of a continuous system is infinite, and each vibration mode has the corresponding natural frequency. For the moving components in power engineering, such as blades, the first-order bending mode is the most common form in actual service. In other aspect, the first-order vibration mode is easy to be conducted in fatigue tests compared to the higher-order vibration modes. Thus only the first-order longitudinal vibration mode has been merely considered in both the conventional electromagnetic loading and ultrasonic loading. Similarly, the first-order bending vibration mode is considered in the present UHF specimen. And the natural angular frequency ω in the first-order mode is the key parameter to be concerned during the design.

$$\omega = f(l, \rho, E, \mu) \tag{1}$$

where ρ, E and μ denote density, Young's modulus and Poisson's ratio, respectively. By dimensional analysis, a derived equation with a dimensionless form can be obtained by transferring Equation (1), as shown in Equation (2):

$$\frac{\omega l}{\sqrt{\frac{E}{\rho}}} = f(\mu) \tag{2}$$

For two specimens with the same shape, material and boundary condition, the only difference is the characteristic length or size l, thus the proportional relation expressed by Equation (3) can be obtained.

$$\frac{\omega_1}{\omega_2} = \frac{l_2}{l_1} \tag{3}$$

Accordingly, any higher natural frequency of a fatigue specimen could be achieved by reducing the size proportionally in theory. However, the clamping reliability and the fatigue dangerous zone (i.e., working section of specimen in which the fatigue failure is most likely to occur) should be simultaneously considered, thus a novel UHF fatigue specimen cannot be determined by simply reducing the size proportionally of the existing vibration-based fatigue specimen. In the present study, an iterative method was adopted to obtain the geometry of UHF fatigue specimen, with the flow chart shown in Figure 1. Here, two basic goals should be mentioned: One of them is that the first-order bending natural frequency or resonance frequency f of UHF fatigue specimen should fall within the range 1600 Hz < f < 2000 Hz, which ensure the testing period of 10^9 cycles is less than one week. Of course, the resonance frequency beyond 2000 Hz is also not welcome in order to avoid possible frequency effect. Furthermore, the present frequency range is close to that of the aeroengine blade with the present titanium alloy in order to make the present fatigue testing results more valuable for the blade. The other basic goal is that the maximum stress σ_{max} in the fatigue dangerous zone should be significantly larger than the mean value σ_m in the same zone. And a relation $\sigma_{max} \geq 1.5\sigma_m$ is adopted in the present study.

Accordingly, finite element method (FEM) was employed to determine the geometry of UHF fatigue specimen. A series of FEM models with different geometries was established by a commercially available FEM code ABAQUS (v6.14, Dassault Systemes, Providence, RI, USA). Generally, the maximum stress levels for HCF and VHCF tests are far less than the yield strength of specimens. Thus the present specimen is modeled as a linear-elastic solid, with the fundamental material properties listed in Table 1. Only the first bending vibration mode was considered in the study and the natural frequencies can be obtained by the Lanczos eigensolver integrated in ABAQUS. And the stress distribution on the surface of the specimen during the vibration can be also obtained.

Table 1. Material parameters of TA11 titanium alloy [19].

Material Parameters	Value
Young's modulus E (GPa)	107
Poisson's ratio μ	0.334
Yield strength σ_y (MPa)	930
Density (g/cm^3)	4.37

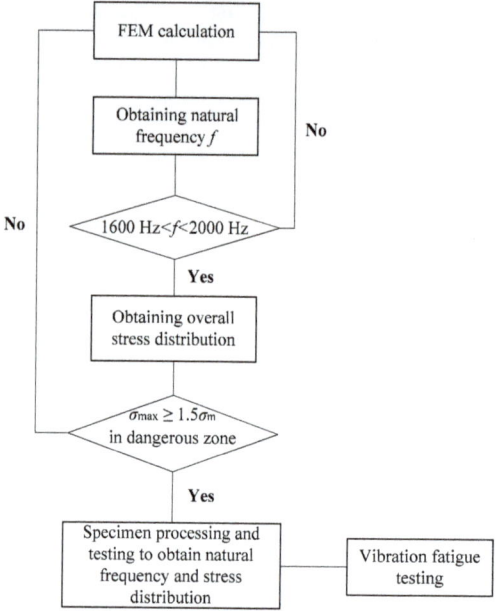

Figure 1. Iterative method to design ultra-high frequency specimen.

After the FEM calculation, a design of UHF specimen was finally determined, with the geometry shown in Figure 2. The two holes in the right side are used to install bolts to mount the specimen on testing system, while the three small holes in the left side are used to adjust the natural frequency and stress distribution of the specimen. The first-order mode bending natural frequency by the FEM calculation is 1775 Hz, which just meets the aforementioned frequency requirement. And the surface normalized axial stress contour S_{11} is shown in Figure 3a. Noting the area with the two mounting holes would be totally clamped by the fixture, thus clamping area can be replaced by the boundary condition of fixed support, and the two holes are not necessarily included in the FEM model.

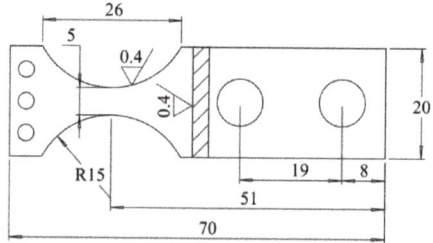

Figure 2. Geometry of the present ultra-high frequency (UHF) fatigue specimen (unit: mm).

In order to validate the FEM result, a stress measurement based on strain gauge was used to obtain the stress values along the central line of the area with arc segment. Benefit from the FEM calculation, a continuous stress distribution curve along the same central path can be obtained and shown in Figure 3b, with O and E representing the origin and end points on the central path, respectively. The comparison of the stress distributions from the FEM and strain gauge can be found in Figure 3b, which shows very good consistency. It should be noted that the maximum stress point is located not exactly at the midpoint, but slight near to the clamping end (normalized location = 0.41353), which is resulted from the bending deformation of the specimen. Furthermore, the mean stress σ_m along the central path can be obtained by the Equation (4):

$$\sigma_m = \frac{1}{l} \int_{path} \sigma_{11} dl \qquad (4)$$

where l denotes the length of the path. The normalized value of σ_m can be hence obtained, with the value of 0.56. Noting the normalized maximum stress σ_{max} is equal to 1, thus the aforementioned relation $\sigma_{max} \geq 1.5\sigma_m$ can be satisfied.

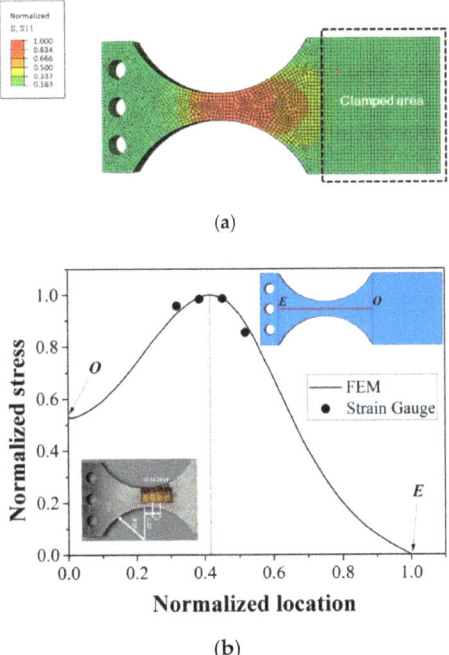

Figure 3. FEM calculation and validation of the present UHF specimen: (**a**) Surface normalized axial stress contour; (**b**) comparison of the stress distributions from FEM and strain gauge (O and E represent the origin and end points on the central path of the specimen).

Furthermore, it should be pointed out the optimal geometry of the present UHF fatigue specimen is dependent on the material and geometrical parameters. And the geometry of the present UHF specimen is proposed based on the present titanium alloy. Although this geometry is not exactly applicable to other materials, it is still important reference geometry for the similar fatigue tests of other materials.

3. Material and Experimental Details

3.1. Experimental Material

A near-alpha titanium alloy TA11 alloy equivalent to Ti-8Al-1Mo-1V was used in the present fatigue study. It has been usually used in advanced turbine engines as low pressure compressor blades and due to its excellent damping capacity, low density, high Young's modulus, and fine welding and anti-oxidation performance [20]. The chemical composition of TA11 used in the present study is list in Table 2.

Table 2. Chemical composition of TA11 titanium alloy (mass fraction/%).

Al	V	Mo	Fe	C	N	H	O	Ti
7.79	1.00	0.98	0.04	0.01	<0.01	0.006	0.06	Balance

3.2. UHF Fatigue Testing Setup

A vibration-based bending fatigue test was subsequently conducted using the present UHF specimens shown in Figure 2. All the specimens were cut from the same batch of raw material in order to make the results reliable. The test was conducted on a vibration-based fatigue testing system, of which the major body is an electrodynamic shaker (ES-10D-240 Electrodynamic Shaker System) located in a soundproof room. The maximum loading capacity of the shaker is 10 kN and the frequency range is 5 to 3000 Hz, which meets the mode-I bending vibration testing requirement of the present specimens. Similar to conventional fully-reversed bending fatigue tests, the ratio between the maximum and minimum stress in the present test is equal to −1.

In order to clamp the specimen firmly, a specified fixture was designed and manufactured. As same as illustrated in Figure 4, one end was clamped firmly and the other end was free, which is shown in Figure 4a. An accelerometer was used to monitor the shaker input load and a laser vibrometer was located above the free end to monitor the amplitude. The excitation direction was vertical to the specimen, with the excitation force having a sine waveform. Simultaneously, a small-size strain gauge was mounted longitudinally on the surface of the specimen, exactly locating at the maximum stress location, as shown in Figure 4a. It should be mentioned that a small-size strain gauge (sensitive pattern area: 1 × 1 mm^2) was adopted since both of the specimen and the fatigue dangerous zone are small.

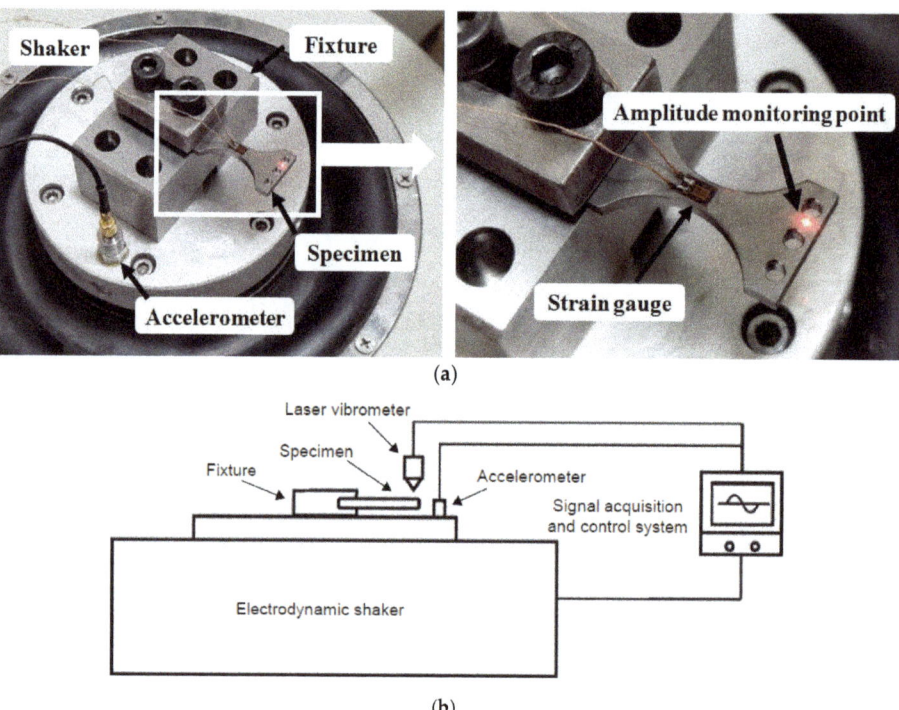

Figure 4. Vibration-based fatigue experiment for the UHF specimen: (**a**) Testing equipment (**b**) Sketch of the experimental system.

In this test, the amplitude of the free edge was a main object feedback, by which the excitation frequency could be automatically adjusted. The specimens were expected to be tested in the resonance condition and the excitation frequency could be adjusted automatically to keep the vibration amplitude stable [16]. Consequently, the amplitude could be steadily controlled. Benefiting from the automatic self-adjusting, the experimental system could run by the means of so-called closed-loop control, which is sketched in Figure 4b.

3.3. Resonance Frequency and Stress Calibration

It is widely known that the response amplitude of a specimen can reach its maximum value when the specimen vibrates under resonance condition for the same excitation load. Therefore, the vibration-based fatigue tests are always expected to be carried out under the resonance condition for the purpose of reducing the power-consuming of the experimental system. Accordingly, the resonance frequency should be identified before the vibration-based fatigue test. In order to obtain the vibration characteristics, the excitations with a series of increasing frequency was imposed upon the specimen, and the excitation frequency-response curves can be hence obtained and the resonant frequency of the specimen can be further determined, with the value of around 1756 Hz, which is obtained by the frequency sweeping testing shown in Figure 5. Noting that specimen is clamped by the designed fixture in the test, the resonant frequency obtained from the test is actually that of the combination of the specimen and the fixture, and it would be less than the natural frequency (i.e., 1775 Hz) of the single specimen, which is obtained by the computation presented in Section 2.

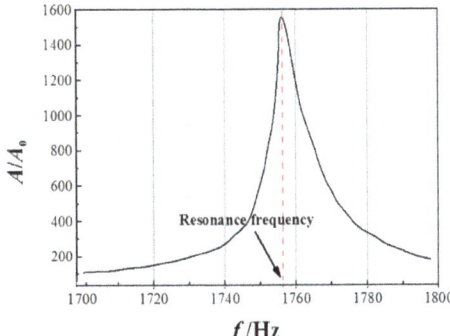

Figure 5. Determination of the resonance frequency of the UHF specimen by frequency sweeping.

During the vibration-based fatigue test process, the resonant frequency was monitored to determine the failure moment of the specimen. In this study, the excitation frequency was preset using the same value of the obtained resonant frequency, which is a stable value (i.e., around 1756 Hz) during the fatigue testing process. As the crack propagates in the fatigue dangerous zone, the resonant frequency decreases gradually to a critical value. When the resonant frequency drop rate reaches to a critical value with the value of 1%, the fatigue test will be terminated and the specimen is considered to be failure.

Since the strain gauge would fail soon after several cycles in the fatigue test, the stress-control of vibration-based fatigue tests has been always achieved by controlling the amplitude. Thus a calibration relation between the measured strain and the amplitude should be determined prior to the fatigue testing. The clamped specimen is similar to a normal slender cantilever beam and the transverse stress can be ignored. Accordingly, three typical values of the amplitude were selected and the peak-valley values of the strain along the 1-direction were measured during the vibration testing. A linear calibration relation between the measured strain (peak-valley value ε_{P_V}) and double-amplitude $2a$ of the present UHF specimen can be obtained, shown in Figure 6. For a specific amplitude a, the value of σ_1 can be gained by the strain gauge measure together with the stress-strain relation $\sigma_1 = 0.5 \cdot E\varepsilon_{P\_V}$.

Thus the present stress-control vibration-based fatigue test is actually the realized by control the amplitude, and various stress levels can be realized by varying amplitude a.

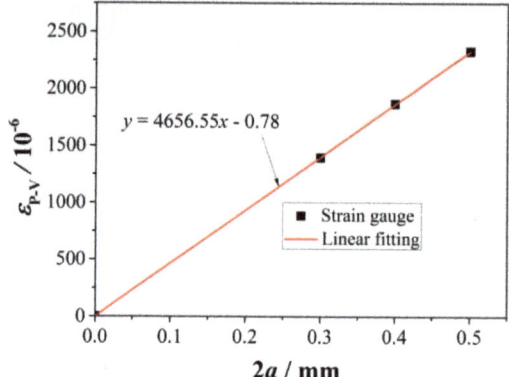

Figure 6. Calibration relation between the measured strain and displacement amplitude of the UHF specimen.

3.4. Fatigue Tests for Comparison

In order to verify and compare the present UHF fatigue testing results, some comparative tests have been conducted from two aspects:

On one hand, the effect of the loading frequency on the testing results should be verified. Some previous studies for VHCF tests have also considered this issue [7,9], But the comparative studies have been always performed among different types of fatigue tests, such as the comparison between the rotating bending fatigue test and ultrasonic fatigue test. Although the loading frequencies of the tests are definitely different, the loading condition also influences the testing results. Accordingly, the effect of the loading frequency can best be considered separately. In the present study, a conventional vibration fatigue (CVF) specimen shown in Figure 7 was used to explore the influence of the loading frequency. The two holes in the right side are used for mounting bolts to fix the specimen on the testing system. The minimum width of the fatigue dangerous zone is 10 mm. The geometry of the conventional vibration fatigue specimen is taken from a Chinese testing standard HB 5277, which is widely used in the field of vibration-based fatigue testing for the aeroengine blade materials in China. After a similar frequency-response test mentioned by Section 3.3, the actual loading frequency close to the resonance frequency is obtained, with the value of approximately 240 Hz.

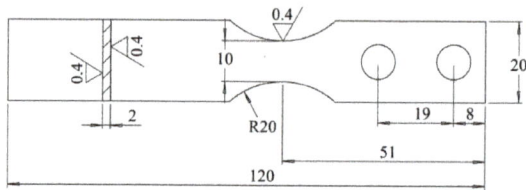

Figure 7. Geometry of the conventional vibration fatigue (CVF) specimen (unit: mm).

One the other hand, the effect of the fatigue loading types should be considered. Some other types of fatigue tests were also conducted, which includes conventional axial loading (CA) fatigue test, rotating bending (RB) fatigue test and ultrasonic axial loading (UA) fatigue test. All these tests have been widely carried out in fatigue community, thus the comparison with them is helpful to verify the applicability of the present testing method in VHCF testing. Here, the CA loading fatigue test and RB fatigue test were carried out in an electromagnetic resonant fatigue testing system and a rotating bending fatigue tester, respectively. And the UA fatigue test was carried out in a commercial ultrasonic fatigue test machine (USF-2000, Shimadzu, Japan). All the fatigue tests were conducted in room temperature, with the stress ratio R equal to -1. The loading frequencies for the CA, RB and UA fatigue tests were 120 Hz, 83.3 Hz and 20 kHz, respectively. Considering the ultra-high loading frequency (i.e., 20 kHz) in the UA fatigue testing, a compressive dry air cooling system was used to cool the UA specimen during the testing in order to ensure the specimen temperature is maintained at room temperature. Furthermore, an infrared thermometer was used to monitor the surface temperature of the specimen during the UA fatigue test.

The geometry of the specimens for comparison is shown in Figure 8. All the specimens have hourglass-type shape. It should be pointed out all the specimens for comparison were machined from the same batch of raw materials with the present UHF specimen, in order to make the testing results more comparable.

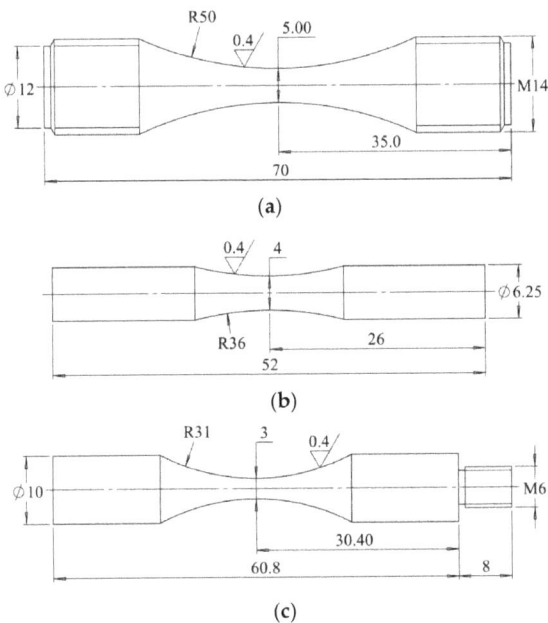

Figure 8. Geometry of the specimens used in fatigue tests for comparison (unit: mm): (**a**) Conventional axial loading (CA) specimen; (**b**) rotating bending (RB) specimen and (**c**) ultrasonic axial loading (UA) specimen.

4. Results and Discussion

4.1. Vibration-Based Fatigue Testing Results

Figure 9 shows the obtained S-N data and the relevant fitting curves for the present vibration-based fatigue tests, which involves the UHF and CVF specimens, with the actual loading frequencies of about 1756 Hz and 240 Hz, respectively. Several stress levels have been selected in the present UHF fatigue testing, covering the stress range from 400 MPa to 540 MPa. And the maximum failure cycle can reach up close to 10^9. In order to compare the results between the two types of vibration-based fatigue tests, the same stress levels have been also considered in the test for CVF specimens. Noting the actual loading frequency of CVF specimens is about 240 Hz, the fatigue testing in the VHCF regime (i.e., >10^7 cycles) has not been considered due to its high time-consuming. Finally, there are 20 and 17 valid data obtained in the UHF and CVF tests, respectively, which are given in Table 3. Considering the significant scatter of the obtained fatigue lives by the tests, several specimens were tested at the same stress level in order to make the results statistically reliable.

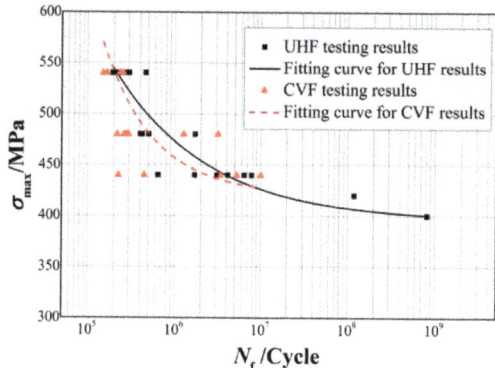

Figure 9. *S-N* data and the fitting curves for the present vibration-based fatigue tests involving the UHF and CVF specimens.

Table 3. *S-N* testing data from the UHF specimens and CVF specimens.

Specimen No.	Stress Level (MPa)	Failure Cycle	Specimen No.	Stress Level (MPa)	Failure Cycle
UHF1	540	4.66×10^5	CVF1	540	1.66×10^5
UHF2	540	2.20×10^5	CVF2	540	2.56×10^5
UHF3	540	2.99×10^5	CVF3	540	1.50×10^5
UHF4	540	2.82×10^5	CVF4	540	2.30×10^5
UHF5	540	2.25×10^5	CVF5	540	2.36×10^5
UHF6	540	1.93×10^5	CVF6	480	1.27×10^6
UHF7	540	2.65×10^5	CVF7	480	2.62×10^5
UHF8	480	4.26×10^5	CVF8	480	2.96×10^5
UHF9	480	4.05×10^5	CVF9	480	3.18×10^6
UHF10	480	5.07×10^5	CVF10	480	2.19×10^5
UHF11	480	4.18×10^5	CVF11	480	2.79×10^5
UHF12	480	1.73×10^6	CVF12	440	5.27×10^6
UHF13	440	6.43×10^6	CVF13	440	1.00×10^7
UHF14	440	1.70×10^6	CVF14	440	3.40×10^6
UHF15	440	7.84×10^6	CVF15	440	1.00×10^7
UHF16	440	3.09×10^6	CVF16	440	4.53×10^5
UHF17	440	4.14×10^6	CVF17	440	2.27×10^5
UHF18	440	6.49×10^5	–	–	–
UHF19	420	1.19×10^8	–	–	–
UHF20	400	8.20×10^8	–	–	–

The *S-N* curves can be obtained by a regression calculation with a three-parameter *S-N* model, which is named as Stromeyer model [21], expressed by:

$$\lg N_f = a - b\lg(\sigma_{max} - S_0) \tag{5}$$

where N_f and σ_{max} denote the failure cycle and stress level (or maximum cyclic stress). And a, b and S_0 are the fitting parameters. Noting σ_{max} would approach to S_0 when N_f approaches to infinite, thus S_0 can be regarded as a fatigue limit from a mathematical point of view. Consequently, the *S-N* curves from Stromeyer model would exhibit obvious curvature characteristics and the fitting model has been widely used in the fatigue community. The values of a, b and S_0 for UHF specimens are 10.7, 2.50 and 395, respectively, while those for CVF specimens are 8.05, 1.32 and 421, respectively.

In addition, typical surface crack morphology when UHF specimen fails are shown in Figure 10. It can be found an obvious continuous crack locating approximately at normalized location of 0.38 initiated at O point shown in Figure 3b, which is close to the maximum stress point determined in Section 2, not at the narrowest width of the specimen. And the crack propagation direction is basically perpendicular to the axial stress direction (i.e., 1-direction). Thus it is reasonable to determine the strain measurement location as mentioned in Section 3.2.

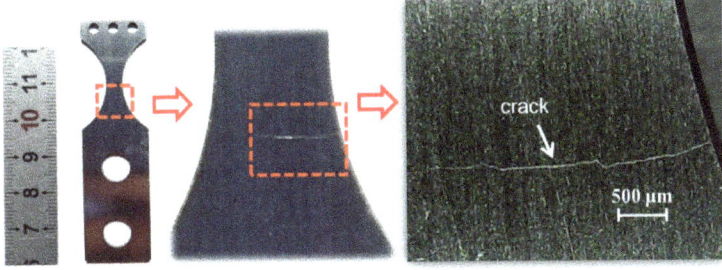

Figure 10. Typical crack morphology when UHF specimen fails.

4.2. Effect of the Loading Frequency

As shown in Figure 9, the S-N curves between the UHF and CVF specimens are close to each other in the same life cycle regime. Both of the UHF and CVF specimens are sheet specimens used in the vibration-based testing with the same type of fatigue loads. The major difference between them is merely the loading frequency, which does not influence the S-N curves clearly as shown in Figure 9. Considering the two cases share the same fatigue stress levels and several testing data have been obtained at those fatigue stress levels, further comparison and analysis can be conducted in order to explore the effect of loading frequency further.

In the present vibration-based fatigue testing, three fatigue stress levels were considered with the values 440 MPa, 480 MPa and 540 MPa. Figure 11 shows the comparison of fatigue lives from the UHF and CVF specimens at these fatigue stress levels. It can be found the fatigue lives for the two types of specimens are close to each other at these stress levels. Strictly speaking, the fatigue lives for the CVF specimen are slightly shorter than those for the UHF specimen, which is clear at the stress level of 440 MPa. One reason leading to the tiny discrepancy is probably the influence of the specimen size. It can be found there is a significant difference in the size of the two types of specimens, despite their shapes are similar. It has been previously pointed out a smaller size fatigue specimen would have a longer fatigue life due to the smaller risky volume and less likelihood of containing defects [22]. Thus the present tiny discrepancy in fatigue lives between the two specimens can be mainly attributed to specimen size instead of the loading frequency.

In addition, the error bars shown in Figure 11 to reflect the dispersion of result data are worth mentioning. The length of the error is generally increased as the fatigue stress level is decreased. It suggests the dispersion of the fatigue lives is more significant for the lower stress level, which has been widely verified by the previous HCF and VHCF researches [23,24]. It should be noted the length of error bars for UHF specimen is significantly shorter than those for CVF at the low fatigue stress levels (i.e., 440 and 480 MPa), which suggests the results for UHF specimens are probably less dispersive than the CVF specimens. Although there is no solid reason to explain it, it is at least concluded the data stability by the present UHF specimens is not inferior to the conventional specimens with the lower frequency.

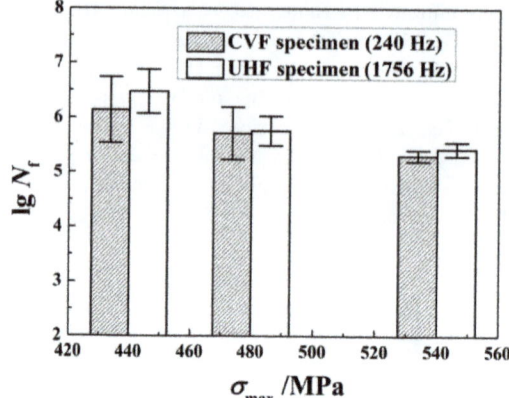

Figure 11. Comparison of fatigue life results from the UHF and CVF specimens for the same fatigue stress levels.

4.3. Effect of the Testing Types

Figure 12 shows the comparison of the results of the present UHF specimens and other types of fatigue specimens in consideration. The S-N curves can be also obtained by Equation (5), with the values of the fitting parameters given in Table 4.

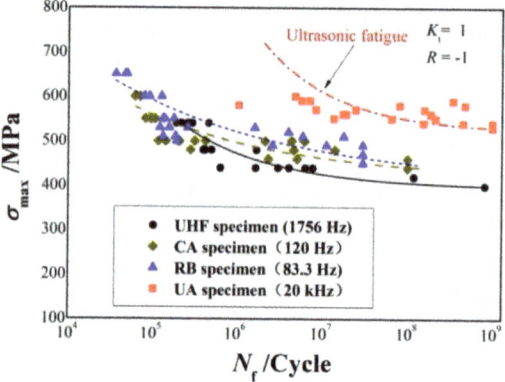

Figure 12. Comparison of the results of ultra high frequency (UHF) specimen and other types of fatigue specimens.

Table 4. Fitting parameters of Equation (5) for the comparative tests.

Testing Type	a	b	S_0/MPa
UA	10.8	1.94	523
RB	13.5	3.97	418
CA	15.8	4.78	407

In general, the S-N data based on the present UHF specimens get close to those based on the conventional HCF methods, including the CA and RB specimens. In contrast, the S-N curve from the UA specimens has significant discrepancy compared with those from all other testing methods. It can be found that the fatigue life data obtained by the UA test are much longer than other types of tests at the same stress level.

Generally, the factors influencing the fatigue testing results includes the temperature rise of specimen, material uniformity and testing method. Noting the cooling system was simultaneously used during the UA testing process, the surface temperature was kept at the range of 12~15°C obtained by the infrared thermometer. Thus the temperature rise of the UA specimens can be neglected. In addition, all the fatigue specimens in this study were obtained based on same batch of Ti-alloy, thus it can be inferred the discrepancy of the results shown in Figure 12 is caused by the testing method.

For fatigue testing method, an important factor is the specimen size, which can be usually considered to explain the discrepancy of the testing data. It has been usually widely thought that the fatigue specimen with smaller size would have longer fatigue life [22,25]. The specimen in the UA testing usually has a small risky volume which means the specimen volume subjected to a stress amplitude larger than the 90% of its maximum value [26]. Some previous studies have attributed the size effect of fatigue specimens to the influence of the risky volume [25,26]. However, another important factor should be noted is the loading frequency. Since both the loading frequency and the risky volume could influence the fatigue life of specimens, the explanation from the risky volume is feasible when the loading-frequencies of the fatigue tests are close to each other. For example, for the present two conventional comparative fatigue tests involving the CA and RB specimens, their loading frequencies are both close to 100 Hz. Accordingly, the discrepancy between their corresponding S-N curves could be explained by the risky volume theory, which means the risky volume of RB specimen is smaller than that of CA, resulting in the fatigue life of RB specimen is longer. It should be pointed out the risky volume of the UA specimen is actually small, but still larger than that of the present UHF and RB specimens. Note the fatigue life obtained by the UA test is much longer than other types of tests, the discrepancy between their results cannot be explained by the risky volume theory.

Instead, another important factor in the testing method is the loading frequency, which could be the major factor to cause the discrepancy of the testing results. It has been found that the fatigue lives of some materials have been proved to be almost unaffected by the loading frequency, while the situation of some other materials are opposite [9]. In general, the materials with an obvious strain rate-related effect are more susceptible to loading frequency [7], thus it can be inferred the present titanium alloy is a material with obvious strain rate-related effect and the UA testing method is probably not suitable for its VHCF testing. In contrast, although the frequency of the present UHF fatigue method is also high (1756 Hz) compared with the conventional testing methods, the obtained fatigue lives by the present UHF method are not clearly influenced by the high frequency.

4.4. Discussion

Although the testing results from the present UHF specimens is generally close to those from CA and RB specimens, especially in the long life regime, the discrepancy between them deserves further explanation from the perspective of the failure criterion. It should be paid attention that the present vibration-based UHF test has a different failure criterion compared with that of the aforementioned conventional fatigue tests. Actually, the failure criterion of the most conventional fatigue tests is the separation of specimen. However, the present UHF test adopts the failure criterion that the resonance frequency of specimens drops by 1% of the initial value. The critical value of resonance frequency used as the failure criterion was determined based on the previous vibration-based fatigue tests [27]. It has been widely known that the life of the crack initiation and the growth of the micro-structurally small crack accounts for a large proportion of the total life in long life regime [28]. For the low stress cases in present UHF test, once the crack initiation occurs, the specimen would fail very soon since the loading frequency is very high. Thus the critical frequency for the failure criterion can be feasible. However, for the high stress cases in the present UHF test, the proportion of crack initiation life in the total life decreases while the proportion of the crack propagation life increases. Consequently, the specimen would not fail very soon after the crack initiation and the growth of the micro-structurally small crack.

Here, it should be clarified the reason why the specimen separation is not applicable for the failure criterion of the present UHF test. Firstly, the present vibration-based fatigue specimen has been usually clamped at only one side, with the other side free. If the testing continues until the specimen is separated, the free side is likely to impact and damage the surrounding devices and personnel when the separation occurs, because the loading frequency is very high. Secondly, the stress control of the testing is achieved by controlling the vibration amplitude of the specimen. As the fatigue test continues, the damage evolution would continue so that the resonance frequency would drop simultaneously. But the vibration amplitude should be maintained during the whole testing period to satisfy the testing stress condition, which needs a close-loop control method. However, in the period before the specimen's final separation, the resonance frequency drops dramatically as the damage evolution develops sharply. Thus it is so difficult to maintain the vibration amplitude at a constant in the final period. Considering the final period approaching to the separation of specimen is usually short, thus a critical frequency drop has been usually adopted as the failure criterion, instead of the specimen separation.

However, the failure criterion that the resonance frequency of specimens drops by 1% is just an empirical one, which has been proved feasible in some conventional vibration-based fatigue tests. Thus it has been adopted as a recommended failure criterion in the Chinese vibration-based fatigue testing standard HB 5277, which is also followed in the present study. However, the failure criterion of the frequency drop by 1% may be not the optimal one for the present non-standard UHF specimen especially in the higher fatigue stress cases. Thus it is necessary to figure out the optimal failure criterion for the present UHF specimen in the future, which helps to further improve the accuracy of the present testing results.

5. Conclusions

In summary, this paper proposes an ultra-high frequency (UHF) fatigue test of a titanium alloy TA11 based on electrodynamic shaker in order to develop a feasible testing method in the VHCF regime. Firstly, a type of UHF fatigue specimen is designed to make its actual testing frequency reach as high as 1756 Hz. Then the influences of the loading frequency and loading types on the testing results of the fatigue life are considered separately, and a series of comparative fatigue tests are hence conducted. The results show the testing data from the present UHF fatigue specimen agree well with those from the conventional vibration fatigue specimen with the loading frequency of 240 Hz. Furthermore, the present UHF testing data show good consistency with those from the axial-loading fatigue and rotating bending fatigue tests. But the fatigue life obtained from the ultrasonic fatigue test is significantly higher than all other fatigue testing results. Thus the proposed ultra-high frequency vibration-based fatigue test will have a good application prospect in the VHCF testing due to its balance of high efficiency and similarity with the conventional testing results.

Author Contributions: Conceptualization, W.X.; Funding acquisition, Y.H. and C.T.; Investigation, W.X.; Methodology, B.Z.; Project administration, H.Y.; Software, B.Z.; Supervision, H.Y., and Y.H.; Validation, X.C; Visualization, Y.Z.; Writing—original draft, W.X. and X.C.; Writing—review & editing, Y.Z. All authors have read and agreed to the published version of the manuscript.

Funding: This work was supported by the National Natural Science Foundation of China (91860112), the National Key Research and Development Program of China (2017YFB0702004), the Materials Special Project (JPPT-KF2008-6-1) and the Open Project from State Key Laboratory of Structural Analysis of Industrial Equipment of DLUT (GZ18116).

Acknowledgments: W.X. would like to acknowledge the advice of Sun Chengqi from Institute of Mechanics, Chinese Academy of Sciences.

Conflicts of Interest: The authors declare no conflict of interest.

References

1. Cervellon, A.; Cormier, J.; Mauget, F.; Hervier, Z. VHCF life evolution after microstructure degradation of a Ni-based single crystal superalloy. *Int. J. Fatigue* **2017**, *104*, 251–262. [CrossRef]
2. Stanzl-Tschegg, S. Very high cycle fatigue measuring techniques. *Int. J. Fatigue* **2014**, *60*, 2–17. [CrossRef]

3. Mayer, H. Recent developments in ultrasonic fatigue. *Fatigue Fract. Eng. Mater. Struct.* **2016**, *39*, 3–29. [CrossRef]
4. Sharma, A.; Oh, M.C.; Ahn, B. Recent Advances in Very High Cycle Fatigue Behavior of Metals and Alloys—A Review. *Metals* **2020**, *10*, 1200. [CrossRef]
5. Nikitin, A.; Palin-Luc, T.; Shanyavskiy, A. Crack initiation in VHCF regime on forged titanium alloy under tensile and torsion loading modes. *Int. J. Fatigue* **2016**, *93*, 318–325. [CrossRef]
6. Huang, Z.Y.; Liu, H.Q.; Wang, H.M.; Wagner, D.; Khan, M.K.; Wang, Q.Y. Effect of stress ratio on VHCF behavior for a compressor blade titanium alloy. *Int. J. Fatigue* **2016**, *93*, 232–237. [CrossRef]
7. Hu, Y.; Sun, C.; Xie, J.; Hong, Y. Effects of Loading Frequency and Loading Type on High-Cycle and Very-High-Cycle Fatigue of a High-Strength Steel. *Materials* **2018**, *11*, 1456. [CrossRef]
8. Morrissey, R.; Nicholas, T. Fatigue strength of Ti–6Al–4V at very long lives. *Int. J. Fatigue* **2005**, *27*, 1608–1612. [CrossRef]
9. Guennec, B.; Ueno, A.; Sakai, T.; Takanashi, M.; Itabashi, Y. Effect of the loading frequency on fatigue properties of JIS S15C low carbon steel and some discussions based on micro-plasticity behavior. *Int. J. Fatigue* **2014**, *66*, 29–38. [CrossRef]
10. Witek, L. Simulation of crack growth in the compressor blade subjected to resonant vibration using hybrid method. *Eng. Fail. Anal.* **2015**, *49*, 57–66. [CrossRef]
11. Cowles, B. High cycle fatigue in aircraft gas turbines—An industry perspective. *Int. J. Fract.* **1996**, *80*, 147–163. [CrossRef]
12. Amabili, M. Geometrically nonlinear vibrations of rectangular plates carrying a concentrated mass. *J. Sound Vib.* **2010**, *329*, 4501–4514. [CrossRef]
13. Scott-Emuakpor, O.; Shen, M.H.H.; George, T.; Cross, C.J.; Calcaterra, J. Development of an Improved High Cycle Fatigue Criterion. *J. Eng. Gas Turbines Power* **2007**, *129*, 162. [CrossRef]
14. Toni Liong, R.; Proppe, C. Application of the cohesive zone model for the evaluation of stiffness losses in a rotor with a transverse breathing crack. *J. Sound Vib.* **2013**, *332*, 2098–2110. [CrossRef]
15. Magi, F.; Di Maio, D.; Sever, I. Damage initiation and structural degradation through resonance vibration: Application to composite laminates in fatigue. *Compos. Sci. Technol.* **2016**, *132*, 47–56. [CrossRef]
16. Yun, G.J.; Abdullah, A.B.M.; Binienda, W. Development of a Closed-Loop High-Cycle Resonant Fatigue Testing System. *Exp. Mech.* **2011**, *52*, 275–288. [CrossRef]
17. Xu, W.; Yang, X.; Zhong, B.; Guo, G.; Liu, L.; Tao, C. Multiaxial fatigue investigation of titanium alloy annular discs by a vibration-based fatigue test. *Int. J. Fatigue* **2017**, *95*, 29–37. [CrossRef]
18. George, T.J.; Seidt, J.; Shen, M.-H.H.; Nicholas, T.; Cross, C.J. Development of a novel vibration-based fatigue testing methodology. *Int. J. Fatigue* **2004**, *26*, 477–486. [CrossRef]
19. Yu, H.; Wu, X. *Materials Data Handbook for Aircraft Engine Design (Part V)*; Aviation Industry Press: Beijing, China, 2014.
20. Yang, K.; He, C.; Huang, Q.; Huang, Z.Y.; Wang, C.; Wang, Q.; Liu, Y.J.; Zhong, B. Very high cycle fatigue behaviors of a turbine engine blade alloy at various stress ratios. *Int. J. Fatigue* **2017**, *99*, 35–43. [CrossRef]
21. Stromeyer, C.E. The Determination of Fatigue Limits under Alternating Stress Conditions. *Proc. R. Soc. Lond. Ser. A Contain. Pap. Math. Phys. Character* **1914**, *90*, 411–425.
22. Paolino, D.S.; Tridello, A.; Chiandussi, G.; Rossetto, M. On specimen design for size effect evaluation in ultrasonic gigacycle fatigue testing. *Fatigue Fract. Eng. Mater. Struct.* **2014**, *37*, 570–579. [CrossRef]
23. Shimizu, S.; Tosha, K.; Tsuchiya, K. New data analysis of probabilistic stress-life (P–S–N) curve and its application for structural materials. *Int. J. Fatigue* **2010**, *32*, 565–575. [CrossRef]
24. Bomas, H.; Burkart, K.; Zoch, H.-W. Evaluation of S–N curves with more than one failure mode. *Int. J. Fatigue* **2011**, *33*, 19–22. [CrossRef]
25. Tridello, A.; Paolino, D.S.; Chiandussi, G.; Rossetto, M. Gaussian specimens for VHCF tests: Analytical prediction of damping effects. *Int. J. Fatigue* **2016**, *83*, 36–41. [CrossRef]
26. Sun, C.; Song, Q. A Method for Predicting the Effects of Specimen Geometry and Loading Condition on Fatigue Strength. *Metals* **2018**, *8*, 811. [CrossRef]
27. Xu, W.; Yang, X.; Zhong, B.; He, Y.; Tao, C. Failure criterion of titanium alloy irregular sheet specimens for vibration-based bending fatigue testing. *Eng. Fract. Mech.* **2018**, *195*, 44–56. [CrossRef]

28. Przybyla, C.P.; Musinski, W.D.; Castelluccio, G.M.; McDowell, D.L. Microstructure-sensitive HCF and VHCF simulations. *Int. J. Fatigue* **2013**, *57*, 9–27. [CrossRef]

Publisher's Note: MDPI stays neutral with regard to jurisdictional claims in published maps and institutional affiliations.

 © 2020 by the authors. Licensee MDPI, Basel, Switzerland. This article is an open access article distributed under the terms and conditions of the Creative Commons Attribution (CC BY) license (http://creativecommons.org/licenses/by/4.0/).

Article

Hydrogen Assisted Fracture of 30MnB5 High Strength Steel: A Case Study

Garikoitz Artola [1,*] and Javier Aldazabal [2]

1. Azterlan, Research and Development of Metallurgical Processes, Aliendalde Auzunea 6, 48200 Durango, Spain
2. Tecnun, University of Navarra, Manuel de Lardizábal 15, 20018 San Sebastian, Spain; jaldazabal@tecnun.es
* Correspondence: gartola@azterlan.es; Tel.: +34-946-21-54-70

Received: 6 November 2020; Accepted: 26 November 2020; Published: 30 November 2020

Abstract: When steel components fail in service due to the intervention of hydrogen assisted cracking, discussion of the root cause arises. The failure is frequently blamed on component design, working conditions, the manufacturing process, or the raw material. This work studies the influence of quench and tempering and hot-dip galvanizing on the hydrogen embrittlement behavior of a high strength steel. Slow strain rate tensile testing has been employed to assess this influence. Two sets of specimens have been tested, both in air and immersed in synthetic seawater, at three process steps: in the delivery condition of the raw material, after heat treatment and after heat treatment plus hot-dip galvanizing. One of the specimen sets has been tested without further manipulation and the other set has been tested after applying a hydrogen effusion treatment. The outcome, for this case study, is that fracture risk issues only arise due to hydrogen re-embrittlement in wet service.

Keywords: hydrogen re-embrittlement; environmentally assisted cracking; galvanic protection; high strength steel

1. Introduction

High strength steels offer multiple design and cost advantages. Their most publicized application is in automotive components, where their use is being promoted by progressively more restrictive CO_2 emission control policies. Components manufactured in these steels favor a reduction in emissions and an improved gas mileage thanks to their lightweight design. Other sectors such as oil & gas are also prone to take advantage of high strength steels, such as in jack-ups and mooring chains for offshore platforms [1]. In this case, weight reduction is relevant to optimize the cost of keeping the structures floating and moored at their intended position.

The fastener market also benefits from high strength steels in terms of cost competitiveness [2]. Employing class 10.9 instead of class 8.8 bolts not only allows a reduction in the number of elements employed in a joint thanks to the 20% higher strength of the 10.9 class; it also means a bolt diameter reduction that is accompanied by a flange size reduction and a reduced installation time. Figure 1 shows a scheme in terms of cost.

Despite the fact that high strength steels are attractive cost-wise, they suffer some drawbacks, such as their increased risk of showing Hydrogen Embrittlement (HE) issues. HE causes a deterioration of mechanical properties which is often related to corrosion processes [3,4] and HE affected components fracture under applied stresses which are well below their design specifications. Bolts are a representative example of components which are concerned by this HE susceptibility, as recognized by the existence of fastener-specific standards to account for HE control in the final product [5].

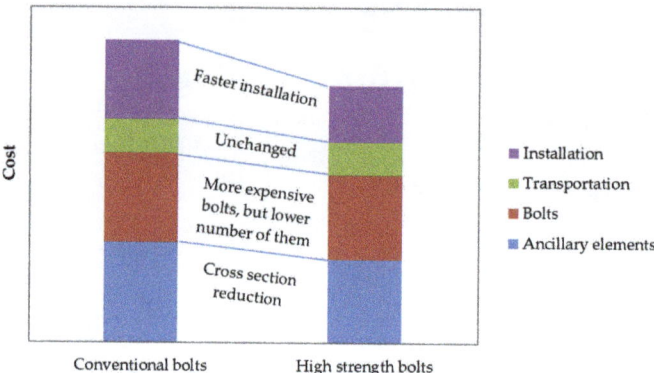

Figure 1. Scheme of the cost advantages that the use of high strength steels implies in bolted joints [2].

When the hydrogen is incorporated into the steels, during the manufacturing process of the component or during service, is referred as internal or external hydrogen, respectively [6].

Internal hydrogen can be incorporated at different steps of the manufacturing route [7], starting upstream in the molten metal [8]. Hydrogen intake is minimized at this stage by applying vacuum degassing techniques on the melt. The importance of this degassing is such that it has become compulsory in some industries [9]. In the case of cast steel parts, the hydrogen left in the melt is expelled from the metal into micro-shrinkage and gas porosity voids during solidifications. The hydrogen trapped in this way recombines into gaseous H_2 in the pores and it is difficult to dissociate it again into atomic hydrogen for removal. In the case of wrought products, hot forging and/or hot rolling aid in the closing the porosity due to solidification and help the removal of hydrogen excess from the steel by the combination of deformation and temperature. The higher the rolling reduction, the lower the hydrogen content in the final material.

After casting and/or forging processes, there are two known sources of internal hydrogen: the intake during heat treatment [10,11], especially when involving austenitizing processes, as hydrogen solubility increases with temperature and in the presence of austenite (Figure 2); and absorption from an electrolyte [12], like acidic media [13] from cleaning and pickling processes or from electrolytic coating processes as in zinc plating [14].

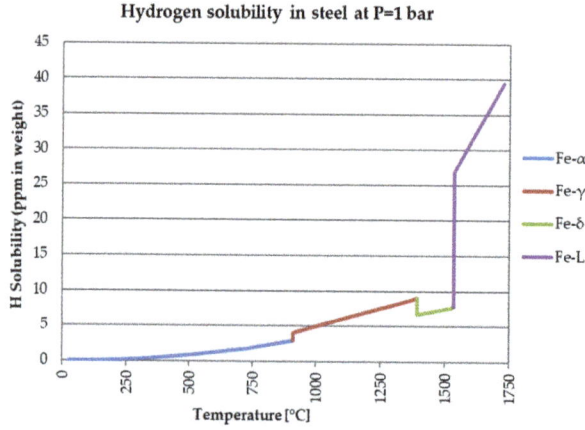

Figure 2. Solubility of H in iron at P-1 bar (made from [15]).

External hydrogen can be incorporated into the steel when the steel is working under high H_2 partial pressures [16] or under acidic conditions, among which sour service with H_2S stands out and has led to the existence of a very specific regulation [17]. A more common source of external hydrogen is the intake from an electrolyte both when impressed current cathodic protection systems and sacrificial coatings are used (e.g., zinc plating or hot dip galvanizing) [18,19]. These processes involving external hydrogen leading to what is known as Environmentally Assisted Cracking (EAC).

Figure 3 shows free corrosion and how galvanic protection processes lead to EAC, in the case of steels working in wet conditions. In free corrosion, iron dissolves in the water as Fe^{2+} ion donating two free electrons. These electrons recombine involving H^+, OH^- and H_2O. When there is an oxygen excess Fe_2O_3 forms on the steel surface. The reduction of H^+ to atomic H does not occur and EAC is avoided. For zinc coated steel in wet service with a discontinuity that exposes the steel substrate, the galvanic potential between the zinc anode and the steel cathodic zone is negative enough to allow hydrogen ion reduction, and drives a H^+ current to the steel surface, where atomic hydrogen can undergo the reaction of the H^+ ion with an electron to yield adsorbed molecular H on the steel surface. From there, the hydrogen enters the steel by diffusion.

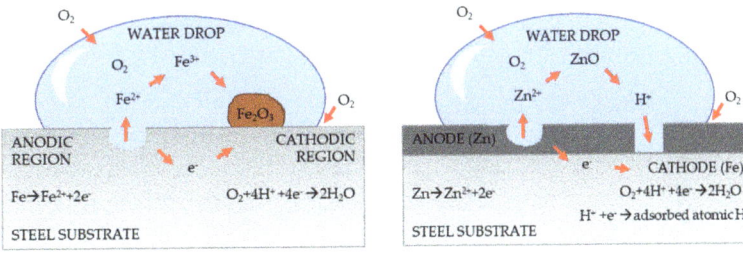

Figure 3. Hydrogen involving reactions in naked (**left**) and zinc coated (**right**) steel in contact with water.

The hydrogen then distributes inside the steel in form of diffusible and trapped hydrogen, the difference being the ability of the hydrogen atoms to move or not across the microstructure [20–22]. The fraction of diffusible hydrogen from the total hydrogen is known to affect EAC, as diffusion allows H atoms to accumulate in the maximum stress triaxiality sites of the material. Thus, hydrogen trap control has become a key resource to develop EAC resistant steels as demonstrated by Fielding [23] and Yamasaki [24]. With a similar approach, as diffusible hydrogen excess tends to effuse from the material when temperature is raised at oven temperatures, oven degassing treatments are industrially used to assess if steel integrity has been compromised by hydrogen [9,25], and this approach is used in this work. More specifically, the Slow Strain Rate Tensile testing (SSRT) results of specimens that have been oven dehydrogenated, compared to non-dehydrogenated, have been used to determine whether internal or external hydrogen was the cause of an in-field failure of a set of galvanized bolts.

The bolt in Figure 4 shows the actual HE case that motivates this study. It is a hot dip galvanized M52 class 10.9 bolt that was installed in an outdoors structure in a coastal onshore location in Northern Spain. The structure consisted of 90 bolts from the same batch and the fractured bolt is part of a set of seven bolts that suffered delayed fracture after a week in service at the same site. All the bolts were tightened with a dynamometric wrench to the desired fastening torque. The torque was calculated to reach the design clamping force of the joint, employing the measured friction coefficient between the bolt and the nut greased threads. The bolt-nut batch had passed regular quality control checks and no deviation was found in the fractured products; they were free of forging defects, decarburization, liquid metal embrittlement or quench cracking.

Figure 4. M52 class 10.9 bolt that failed in service due to delayed fracture.

When the failure surface of the cracked bolts was inspected, a trans-granular brittle cleavage pattern was observed, which is shown in Figure 5a, while the expected fractographic texture for the bolt corresponds to a ductile failure pattern like that shown in Figure 5b.

Figure 5. (a) Brittle cleavage texture in the crack surface of the M50 class 10.9 bolt shown in Figure 1; (b) ductile collapse texture in the tensile specimens extracted from the shaft of the same bolt; (c) microstructure of the bolt in the center of the shaft diameter.

This type of delayed fracture has been reported in hot dip galvanized bolts with sections fully immersed in water [26], EAC being the root cause of failure. The failed bolts of concern in this case study, though, were not immersed in water. Thus, this work is aimed at answering the following question: was the delayed fracture due to HE from internal hydrogen introduced by the manufacturing process or was it due to EAC from external hydrogen taken up in a scenario of water drop condensation on coating discontinuities at the bolt surface?

With this question in mind, a SSRT based study was performed with the same material supply of fractured bolts. Tests on dehydrogenated and non-dehydrogenated samples were performed at different stages of the industrial production process (delivery condition, heat treatment and galvanizing) and, for the galvanized specimens were tested, both immersed in seawater and in dry conditions.

2. Materials and Methods

The material employed for the study is steel grade 30MnB5 according to UNE-EN 10083-3 [27] in bright bar format according to UNE-EN 10263-4 [28]. The material was received with dimensions of ϕ48 mm × 450 mm length (Figure 6) and in the usual condition of raw material for the manufacture of bolts. This material was sourced from the same mill as the fractured M52 bolts that motivated this study.

Figure 6. 30MnB5 bars employed for the study.

The chemical composition of the material was checked during incoming inspection. Table 1 shows the result of the analyses performed by spark spectrometry, with an expanded measurement uncertainty (K = 2) of ±0.03% for all the elements of the table. It is noted that the material meets the specifications set for alloyed quench and tempering steels under the designation 30MnB5, except for a slight excess in Mn content. The observed addition of Cr is intended to improve hardenability and is in accordance with the limits imposed by the current regulations [27], which allow the 30MnB5 nomenclature to be maintained.

Table 1. Chemical composition check (by weight%) of 30MnB5 bars.

Sample		C	Si	Mn	P	S	Cr	B
Specimens *		0.29	0.24	1.49	<0.015	<0.005	0.50	0.0026
Specification 30MnB5	Min.	0.27	-	1.15	-	-	-	0.0008
	Max.	0.33	0.40	1.45	0.025	0.035	**	0.0050
Class requirements 8.8/9.8/10.9	Min.	0.20	-	-	-	-	-	-
	Max.	0.55	-	-	0.025	0.025	-	0.003

* Results of the chemical analysis performed on the bars by spark emission spectroscopy. ** Additions up to 2% Cr permissible by standard [27] for improved hardenability.

The material also meets the chemical composition requirements established for Class 8.8 to 10.9 [29] fasteners. It should be remembered that the bolt classes are determined mainly by the resistance levels as shown in Table 2.

Table 2. Nominal values of mechanical properties for high-strength bolt classes [29].

Class	Yield Strength $R_{p0.2}$ (MPa)	Ultimate Tensile Strength R_m (MPa)	Elongation E (%)	Reduction in Area RA (%)
8.8	>640	>800	>12	>52
9.8	>720	>900	>10	>48
10.9	>900	>1000	>9	>48
12.9	>1080	>1200	>8	>44

For the study of internal hydrogen uptake during the production process, 27 specimens were prepared and divided into three sets of specimens, one for each stage of production (Figure 7):

- Seven specimens were used to study the supply condition of the raw material.
- Twenty were subjected to a quench and tempering treatment in industrial installations accompanying a 30MnB5 and class 10.9 M32 bolt manufacturing order. Seven of these fourteen were used to study internal hydrogen uptake during heat treatment and three for submerged testing.
- The remaining ten heat treated specimens were hot dip galvanized in industrial installations accompanying the same 30MnB5 and Class 10.9 M32 bolt manufacturing order. Seven were used to study internal hydrogen uptake in the galvanizing process and three for submerged testing.

The quench and tempering sequence that was applied is reported below, and it reproduces that of the fractured M52 bolts that motivated this work (same furnaces and processing conditions):

- Austenitizing in a continuous furnace at a setpoint of 875 °C with a dwell time of 30 min. The carbon activity was controlled in the atmosphere to prevent decarburization.
- Oil quenching.
- Tempering in a continuous furnace set at 540 °C for a dwell time of 120 min followed by air cooling, accompanying the same load industrial load of M32 bolts. The furnace atmosphere was controlled to prevent decarburization.

Non-acidic surface conditioning (cleaning, etching and rinsing) and hot dip galvanizing was performed in an industrial production facility.

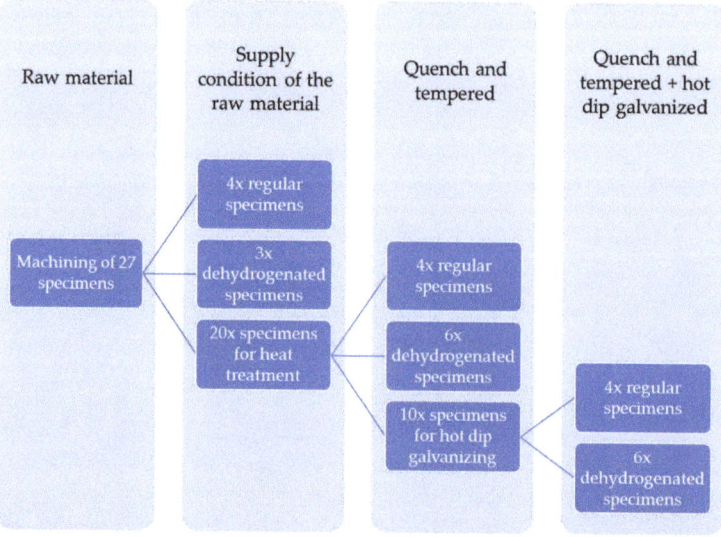

Figure 7. Materials for the study of internal hydrogenation in the manufacturing process for class 10.9 bolts.

Three specimens from each set were submitted to SSRT in air at room temperature without applying a dehydrogenating treatment, while the other three were dehydrogenated in an oven at 250 °C for 2 h before being tested under the same conditions. This baking process at 250 °C is considered sufficient to express a drop in the diffusible hydrogen from the material affected by delayed fracture [9], if any. A loss of ductility of non-conditioned samples relative to dehydrogenated samples would indicate the presence of diffusible hydrogen, incorporated into the material during the production process.

For the heat treated and galvanized specimens, three further specimens were dehydrogenated to perform SSRT immersed in seawater, to assess the severity of the EAC produced by the concurrence of the zinc coating and water. All tests were carried out at room temperature (23 °C). Table 3 summarizes the tests that were performed, indicating the identification code that will be used to refer to each material and testing condition in the following.

Table 3. Summary of the tensile testing battery.

Identification	Description	Conditioning	Testing Media	Type of Test	Repeats
DC	Delivery Condition	Non Dehydrogenated	Air	Regular Tensile	1
DC-ND				SSRT	3
DC-DH		Dehydrogenated		SSRT	3
QT	Quench and Tempered	Non Dehydrogenated	Air	Regular Tensile	1
QT-ND				SSRT	3
QT-DH		Dehydrogenated		SSRT	3
QT-SW			Seawater		3
GA	Quench and tempered + Hot-dip GAlvanized	Non Dehydrogenated	Air	Regular Tensile	1
GA-ND				SSRT	3
GA-DH		Dehydrogenated		SSRT	3
GA-SW			Seawater		3

As is shown in Table 3, one specimen per set was tested under conventional tensile testing conditions [30] to verify that the material performance was as expected and to accept the material condition for further testing. The results of these tests are presented in Table 4. It is noted that the bars in delivery condition could be used directly to manufacture Class 8.8 bolts. After heat treatment, the specimens met the mechanical requirements for class 10.9 bolts as expected. It is important to stress that the heat treatments were carried out in all cases on specimens already machined and not on bars, so that the detection of surface effects caused by the process was not hidden by further machining and it was verified that the control of the protective atmosphere of the furnace had been correct (absence of carburized or decarburized layer).

Table 4. Verification of the mechanical properties of 30MnB5 samples. The tolerances correspond to the expanded measurement uncertainty of each result (K = 2).

Identification		Yield Strength $R_{p0.2}$ (MPa)	Ultimate Tensile Strength R_m (MPa)	Elongation E (%)	Reduction in Area RA (%)
DC		734 ± 7	847 ± 9	13.9 ± 1.5	54 ± 1
QT		1038 ± 10	1100 ± 11	15.1 ± 1.5	67 ± 1
GA		1037 ± 10	1097 ± 11	12.4 ± 1.5	55 ± 1
Specification	Class 8.8	>640	>800	>12	>52
	Class 10.9	>900	>1000	>9	>48

The media used in submerged specimen tests was a seawater substitute according to ASTM D1141 [31]. All electrolytes used for the experimental part of this work were prepared in accordance with this standard in the heavy metal version and with the composition indicated in Table 5. The distilled water used as the base of the mixture met in all preparations the level of purity required by ASTM D1193 [32], Type II. The pH of the electrolyte was adjusted in all cases to 8.2 ± 0.1 by additions of NaOH 0.1N after the compound mixing process was complete.

Table 5. Chemical composition of synthetic seawater employed in the tests.

Compound	Concentration (g/L)
NaCl	24.53
$MgCl_2$	5.20
Na_2SO_4	4.09
$CaCl_2$	1.16
KCl	0.695
$NaHCO_3$	0.201
KBr	0.101
H_3BO_3	0.027
$SrCl_2$	0.025
NaF	0.003
$Ba(NO_3)_2$	0.0000994
$Mn(NO_2)_2$	0.0000340
$Cu(NO_3)_2$	0.0000308
$Zn(NO_3)_2$	0.0000096
$Pb(NO_3)_2$	0.0000066
$AgNO_3$	0.00000049

The SSRT method employed for the assessment of process hydrogenations and zinc coating induced EAC is explained in ASTM G129-00 [33]. This method consists of a test analogous to that of uniaxial tensile testing, but at a very low strain rate. Using a reduced deformation rate (e.g., 10^{-5} s^{-1}) is a key element in the case of studying external embrittlement processes as in this work, as it allows hydrogen distribution to evolve in the microstructure. In conventional tensile test speeds, are at least an order of magnitude faster [30] than in slow strain rate tests, and hydrogen absorption kinetics and diffusion kinetics are not able to modify material behavior and therefore do not allow hydrogen embrittlement to be assessed. It is important to remark that SSRT results are comparative in nature, i.e., they are used to determine whether one material behaves better than another in the same embrittling conditions, or whether the same material has different susceptibility to HE in two different hydrogenation environments. For this reason, the SSRT results are usually provided as a ratio between the values obtained for the condition of interest and a control condition against which it is compared.

The SSRT were performed on a Zwick Roell Model 1475 universal machine (Zwick Roell, Ulm, Germany) with a maximum load capacity of 100 kN. Circular cross-sectional specimens of ϕ10 mm were used according to UNE EN-ISO 6892-1:2017 (Figure 8).

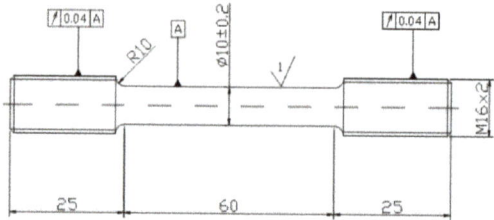

Figure 8. Manufacturing sketches of the specimens used in tests at low deformation speed. Dimensions in mm.

The 30MnB5 steel specimens were obtained from bars of ϕ48 mm as shown in Figure 9. The heat treatments and galvanizing were applied directly to the specimens. In this way, the skin of the tested material reproduces the working condition of interest, like that of the commercial bolts whose failure motivated this work.

The tests were carried out in all cases with a crosshair feed of 0.03 mm/min, which corresponds to a strain rate of 10^{-5} s^{-1}.

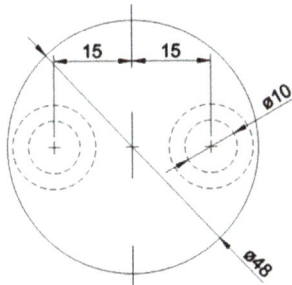

Figure 9. Specimen extraction sketch of bolt steel specimens. Dimensions in mm.

3. Results and Discussion

The average results of the SSRT tests targeted to assessing the HE at different stages of the manufacturing process are shown in Table 6. These tests were performed in air. It should be noted that these averages are calculated on three specimens and, to avoid generating a sense of false accuracy, experimental deviations corresponding to the 95% confidence interval on the average have been included using the expression (1).

$$IC95 = \pm \sqrt{(U(K=2))^2 + \left(\frac{t_{\alpha=0.05,2} \cdot S_v}{\sqrt{3}}\right)^2} \qquad (1)$$

where U (K = 2) is the expanded uncertainty of each individual measure, $t_{\alpha} = 0.05, 2$ is the value of Student's t for a significance value of 0.05% in a distribution of two tails and two degrees of freedom and S_v is the standard deviation.

Table 6. Average values of mechanical properties obtained in SSRT at different stages of the production process of class 10.9 bolts with 30MnB5 steel.

Identification	Yield Strength $R_{p0.2}$ (MPa)	Ultimate Tensile Strength R_m (MPa)	Elongation E (%)	Reduction in Area RA (%)
DC-ND	720 ± 25	841 ± 19	13.7 ± 1.7	51.4 ± 7.9
DC-DH	756 ± 27	850 ± 25	12.9 ± 1.8	52.2 ± 9.6
QT-ND	1022 ± 31	1091 ± 23	14.0 ± 2.5	62.9 ± 6.4
QT-DH	1013 ± 19	1085 ± 18	14.0 ± 1.7	62.5 ± 3.9
GA-ND	1016 ± 36	1087 ± 19	13.4 ± 3.3	58.4 ± 5.4
GA-DH	1007 ± 17	1069 ± 52	13.3 ± 1.6	60.7 ± 1.3

To evaluate if a step in the process had caused any significant change in HE, the results in Table 6 were analyzed by hypothesis tests which were performed for the difference in the averages of normal distributions. The hypothesis tests concerned all the four SSRT outputs considered in this work, $R_{p0.2}$, R_m, E and RA with a significance level of α = 0.01. The conditions that were chosen for the analyses are the following:

- DC-ND versus DC-DH: to assess the possible HE incoming from the hot rolling process.
- QT-ND versus QT-DH: to assess whether there was any HE caused by heat treatment or not.
- GA-ND versus GA-DH: to assess if the galvanizing process caused any HE.

- QT-ND versus GA-ND and QT-DH versus GA-DH: to assess if galvanizing modified the mechanical behavior from the heat-treated condition.

There is no statistically significant difference between the average values of the experimental data in any of the comparisons stated above. Thus, no effect of the production process can be associated with HE.

Figure 10 has been elaborated to help in visualizing this fact. It plots the average values in Table 6 with their standard deviation as the parameters of Normal probability distribution functions. Starting with the values of strength, the results have been divided first by class and then by condition. Figure 10a shows, that for the raw material in delivery condition, R_m distributions overlap by a high percentage, while $R_{p0.2}$ coincide in about 50% of their dispersion. Accounting for the reduced number of repeats, this 50% overlapping has been enough to discard the existence of statistically significant differences between the averages of $R_{p0.2}$ for DC-ND and DC-DH.

The plots in Figure 10b represent the averages for QT and GA specimen sets. These curves are similar to those of DC specimens in the sense that the curves overlap at a high percentage. This fact removes the possibility of affirming that the difference observed between the means is significant, even for the most disperse values such as R_m on GA-DH when working with three repeats.

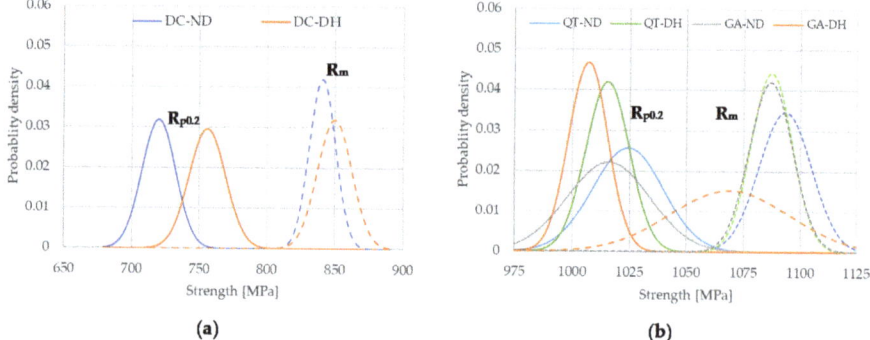

Figure 10. Probability density plot assuming Normal distribution of average strengths and their uncertainties in Table 6 as mean and standard deviation values. (**a**) Plot for the DC Class 8.8 specimens. (**b**) Plot for the QT and GA Class 10.9 specimens.

The same plotting approach for E and RA is shown in Figure 11. For these two properties, which are highly affected by HE and EAC, the curve overlapping is even higher than that for the strengths. This confirms that no major mechanical property decay due to production process-related internal hydrogen HE was produced in the original M52 fractured bolts that motivated this work. Major decays would have been detected even with three repeats.

Regarding the SSRT tests in seawater, targeted to assess the weight of EAC in the failure of the fractured M52 bolts, Table 7 gathers the obtained average results. In this case, the hypothesis tests were again performed in terms of the existence of a difference in the average of the results. The following results were compared:

- QT-DH versus QT-SW: to assess if the immersion in seawater causes any noticeable EAC in the absence of the galvanic protection coating.
- GA-DH versus GA-SW: to assess if the immersion in seawater causes any noticeable EAC in the presence of the galvanic protection coating.

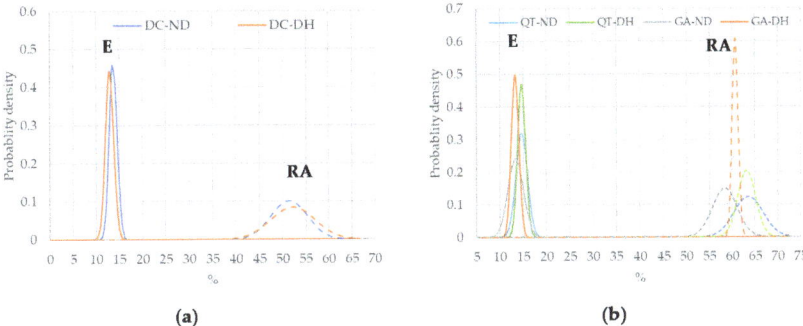

Figure 11. Probability density plot assuming Normal distribution of average elongations and area reduction and their uncertainties in Table 6 as mean and standard deviation values. (**a**) Plot for the DC Class 8.8 specimens. (**b**) Plot for the QT and GA Class 10.9 specimens.

Table 7. Average values of the mechanical properties obtained in water submerged SSRT with 30MnB5 steel treated to class 10.9 bolt strength, both for naked and hot-dip galvanized specimens.

Condition	Yield Strength $R_{p0.2}$ (MPa)	Ultimate Tensile Strength R_m (MPa)	Elongation E (%)	Reduction in Area RA (%)
QT-SW	1014 ± 10	1087 ± 23	13.8 ± 2.6	61.7 ± 4.2
GA-SW	1018 ± 20	1090 ± 26	11.3 ± 3.8	43.6 ± 6.8

The outcome of the statistical analysis indicates that the effect of the immersion is not significant in terms of strength for either of the conditions; neither naked (QT), nor galvanized (GA). Figure 12 reflects this fact, as the gaussian curves overlap clearly for QT (Figure 12a) and slightly less so for GA (Figure 12b) specimens. The most doubtful result is observed for R_m in GA-DH and GA-SW, but again about 50% of the curve overlaps in the worst case and three repeats do not allow the conclusion that this is a significant variation.

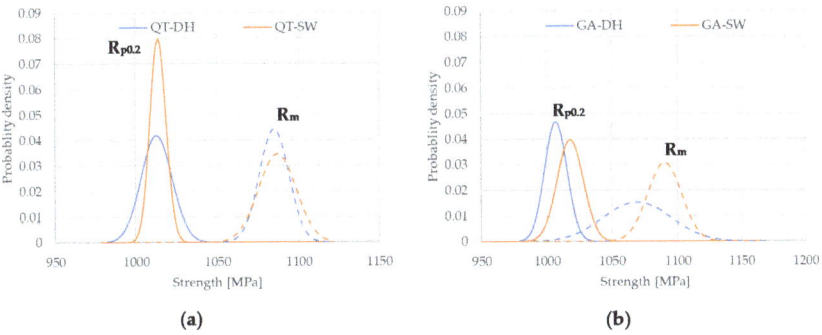

Figure 12. Probability density plot assuming Normal distribution of average strengths and their uncertainties in Table 7 as mean and standard deviation values. (**a**) Plot for the QT specimens. (**b**) Plot for GA specimens.

When the hypothesis tests are performed on the difference of averages of the ductility related properties, the scenario changes, not for the elongation, E, but for the reduction in area, RA. The outcome of the statistical analysis is that the drop in the average RA is significant for the galvanized specimens tested in submerged condition, with a level of significance of $\alpha = 0.01$. Figure 13 plots this: the RA curves of the graph in Figure 13a show that the Gaussian for GA-DH and the Gaussian for GA-SW do not touch each other. Consequently, it is statistically sound to affirm that an embrittlement process has been developed in the GA-SW SSRT specimens and their differences in both E and RA can be considered as an actual effect of EAC and not a statistical artifact. This combination of hot dip galvanizing Zn coatings and water as the critical factor for the failure of bolts in the field has been reported by other case studies [26,34].

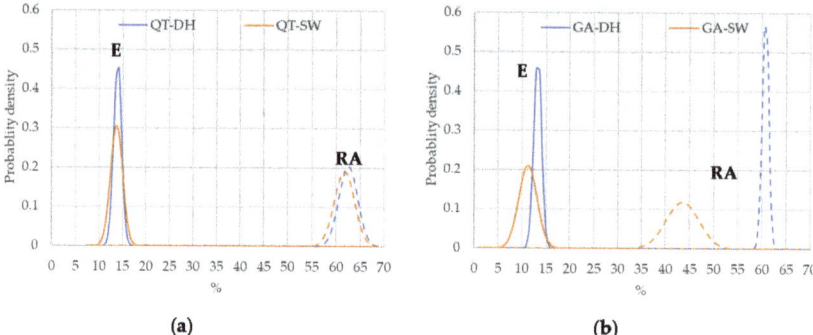

Figure 13. Probability density plot assuming a Normal distribution of average strengths and their uncertainties in Table 7 as mean and standard deviation values. (**a**)Plot for the QT specimens. (**b**) Plot for GA specimens.

Once the presence of EAC is confirmed, it is possible to assess the embrittlement susceptibility of the GA-SW condition, employing the embrittlement ratios as indicated in the standard testing method ASTM G129. The embrittlement ratios measured in this case for the average E and RA values in GA-SW condition using GA-DH as control condition are 85% and 72% respectively (Table 8).

Table 8. Embrittlement susceptibility ratios for the GA-SW condition.

Control Condition	EAC Condition	RE = E_{GA-SW}/E_{GA-DH}	RRA = RA_{GA-SW}/RA_{GA-DH}
GA-DH	GA-SW	85%	72%

The EAC was also confirmed by visual inspection of the fractured SSRT specimens (Figure 14). Only GA-SW specimens showed a very severe cracking pattern all along the necking section. The cracks caused by EAC appear remarked in white, due to material deposition from the electrolyte in the areas of the substrate exposed by the appearance of the cracks themselves. Similar EAC cracking patterns had been observed in previous work [18] for steels with the same strength level as the GA-DH specimens, when a cathodic protection potential was applied to seawater immersed SSRT.

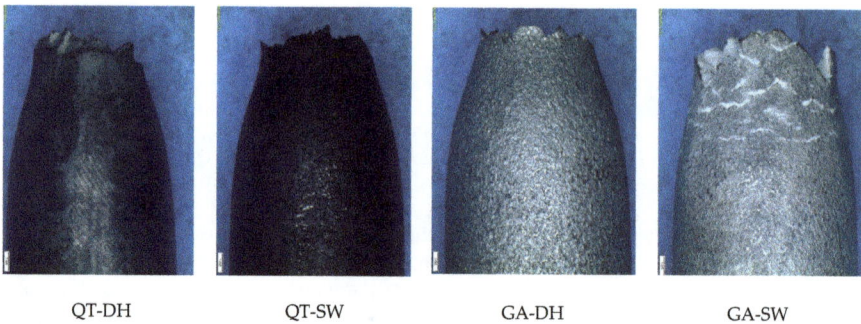

QT-DH   QT-SW   GA-DH   GA-SW

Figure 14. Pictures of the necking areas of representative SSRT specimens showing no EAC except for GA-SW.

4. Conclusions

Tensile testing specimens were manufactured following exactly the same raw material grade and source, and the same industrial manufacturing route (same suppliers and process parameters), as a set of class 10.9 hot-dip galvanized M52 bolts that had suffered a delayed fracture. The intention was to elucidate whether the fracture was HE affected by internal hydrogen uptake of the bolts during their manufacture, or was an EAC issue affected by external hydrogen uptake of the bolts due to the water condensation on a discontinuity of the zinc coating. The results have made clear that the industrial production process does not promote the presence of internal HE in the studied steps: the supply condition of the raw material, the heat treatment, and the hot-dip galvanizing. The specimens processed according to the production route did not require subsequent dehydrogenation processes. Though this does not ensure that there cannot be a mistake in the supply chain, it at least shows the right capability and points to external hydrogen as the principal process involved in the failure of the M52 bolts.

Regarding EAC, it has been observed that mere immersion in seawater does not produce any embrittlement effect on the naked material, while the presence of a galvanic coating in this medium causes a significant susceptibility to EAC. In this particular case study of bolt delayed fracture, the evidence points to a combination of poor zinc coating condition (no matter whether from the galvanizing or due to scratching during poor handling) and the presence of condensation from seawater droplets or aerosol in the installation site (coastal Northern Spain) as the origin of the problem. As the EAC did not affect the whole installed bolt set, it is likely that some of the bolts were unscratched, as the whole batch was installed under the same ambient humidity. Though hydrogen induced fracture-preventing practices are usual in bolt industry, small details such as poor handling can still lead to premature failures. Even in the absence of coating defects, a sufficient coating thickness must be guaranteed, as thin zinc layers may eventually dissolve and expose the EAC susceptible steel substrate.

Author Contributions: Conceptualization, J.A. and G.A.; methodology, J.A. and G.A.; validation, G.A. and J.A.; formal analysis, G.A. and J.A.; writing—original draft preparation, G.A.; writing—review and editing, G.A. and J.A.; supervision, J.A. Both authors have read and agreed to the published version of the manuscript.

Funding: This research received no external funding.

Conflicts of Interest: The authors declare no conflict of interest.

References

1. Billingham, J.; Sharp, J.; Spurrier, J.; Kilgallon, P. *Review of the Performance of High Strength Steels Used Offshore, HSE Books*; TSO: London, UK, 2003.
2. Uno, N.; Kubota, M.; Nagata, M.; Tarui, T.; Kanisawa, H.; Azuma, K.; Miyagawa, T. Super-high-strength bolt "SHBT(R)". In *Nippon Steel Technical Report*; Nippon Steel Corporation: Tokyo, Japan, 2008; Volume 97, pp. 95–104.
3. Milella, P. *Fatigue and Corrosion in Metals*; Springer: Milan, Italy, 2013; pp. 689–729.
4. Anderson, T. *Fracture Mechanics: Fundamentals and Applications*; CRC Press: Boca Raton, FL, USA, 2005.
5. ASTM. *F1940-07a Standard Test Method for Process Control Verification to Prevent Hydrogen Embrittlement in Plated or Coated Fasteners*; American Society for Testing Materials: West Conshohocken, PA, USA, 2014.
6. ASTM. *F1624—12 Standard Test Method for Measurement of Hydrogen Embrittlement Threshold in Steel by the Incremental Step Loading Technique*; American Society for Testing Materials: West Conshohocken, PA, USA, 2012.
7. Bigeev, V.; Nikolaev, A.; Sychkov, A. Effect of the production factors on the hydrogen saturation of steel. *Manuf. Ferr. Met.* **2013**, *6*, 15–21. [CrossRef]
8. Ravinchandar, D.; Balausamy, T.; Gobinath, R.; Balanchandran, G. Behavior of hydrogen in industrial scale steel melts. *Trans. Indian Inst. Metall.* 2018. [CrossRef]
9. DNV-GL. *DNVGL-OS-E302 Offshore Mooring Chain, Offshore Standard*; Det Norske Veritas group: Oslo, Sweden, 2018.

10. Zhang, M.; Wang, M.; Dong, H. Hydrogen absorption and desorption during heat treatment of AISI 4140 steel. *J. Iron Steel Res. Int.* **2014**, *21*, 951–955. [CrossRef]
11. Cho, L.; Sulistiyo, D.; Seo, E.; Jo, K.R.; Kim, S.; Oh, K.; Cho, Y.; De Cooman, B. Hydrogen absorption and embrittlement of ultra-high strength aluminized press hardening steel. *Mater. Sci. Eng. A* **2018**, *734*, 416–426. [CrossRef]
12. El-Yazgi, A.; Hardie, D. The embrittlement of a duplex stainless steel by hydrogen in a variety of environments. *Corros. Sci.* **1996**, *38*, 735–744. [CrossRef]
13. Elhoud, A.; Renton, N.; Deans, W. Hydrogen embrittlement of superduplex stainless steel in acid solution. *Int. J. Hydrogen Energy* **2010**, *35*, 6455–6464. [CrossRef]
14. Hardie, D.; Charles, E.; Lopez, A. Hydrogen embrittlement of high strength pipeline steels. *Corros. Sci.* **2006**, *48*, 4378–4385. [CrossRef]
15. Pasadani, R.; Strauch, D.; Wienkelman, J. Landölt-Borstein: Numerical data and functional relationships in science and technology—Group IV. *Phys. Chem.* **2017**, *5*. [CrossRef]
16. Murakami, Y. The effect of hydrogen on fatigue properties of metals used for fuel cell systems. *Int. J. Fract.* **2006**, *138*, 167–195. [CrossRef]
17. International Organization for Standardization. *ISO 15156-1:2015 Petroleum and Natural Gas Industries—Materials for Use in H2S-Containing Environments in Oil and Gas Industries—Part 1: General Principles for Selection of Cracking-Resistant Materials*; International Organization for Standardization: Geneva, Switzerland, 2015.
18. Artola, G.; Arredondo, A.; Fernández-Calvo, A.; Aldazabal, J. Hydrogen embrittlement susceptibility of R4 and R5 high strength mooring steels in cold and warm seawater. *Metals* **2018**, *8*, 700. [CrossRef]
19. Figueroa, D.; Robinson, M. The effects of sacrificial coating on hydrogen embrittlement and re-embrittlement of ultra-high strength steels. *Corros. Sci.* **2008**, *50*, 1066–1079. [CrossRef]
20. Nagumo, M. *Fundamentals of Hydrogen Embrittlement*; Springer Science + Bussiness Media: Singapore, 2016.
21. Poound, B. Hydrogen trapping in high-strength steels. *Acta Mater.* **1998**, *46*, 5733–5743. [CrossRef]
22. Liu, Y.; Wang, M. Hydrogen trapping in high strength martensitic steel after austenitized at different temperatures. *Int. J. Hydrogen Energy* **2013**, *38*, 14364–14368. [CrossRef]
23. Fielding, L.; Song, E.; Han, D.; Bhadeshia, H.; Suh, D.-W. Hydrogen diffusion and the percolation of austenite in nanostructured bainitic steel. *Proc. R. Soc. A* **2014**, *470*, 20140108. [CrossRef]
24. Yamasaki, S.; Bhadeshia, H. M4C3 precipitation in Fe-C-Mo-V steels and relationship to hydrogen trapping. *Proc. R. Soc. A* **2006**, *462*, 2315–2330. [CrossRef]
25. DNV-GL. *DNVGL-CP-0237 Offshore Mooring Chain and Accessories, Class Programme*; Det Norske Veritas group: Oslo, Sweden, 2018.
26. Chung, Y.; Fulton, L. Environmental hydrogen embrittlement of G4140 and G4340 steel bolting in atmospheric versus immersion services. *J. Fail. Anal.* **2017**, *17*, 330–339. [CrossRef]
27. Spanish Association of Standardization and Certification. *UNE EN 10083-3:2008 Steels for Tempering and Tempering—Part 3: Technical Conditions for the Supply of Alloyed Quality Steels*; AENOR: Madrid, Spain, 2008.
28. Spanish Association of Standardization and Certification. *UNE-EN 10263-4 Bars, Wire Rod and Wire for Cold Deformation and Extrusion. Part 4: Technical Conditions for the Supply of Tempering and Tempering Steels*; AENOR: Madrid, Spain, 2018.
29. Spanish Association for Standardization and Certification. *UNE-EN ISO 898-1 Mechanical Characteristics of Carbon Steel and Alloy Steel Fasteners—Part 1: Bolts, Screws and Bolts with Specified Quality Classes*; AENOR: Madrid, Spain, 2015.
30. Spanish Association of Standardization and Certification. *UNE-EN ISO 6892-1:2017 Metallic Materials. Traction Tests. Part 1: Test Method at Room Temperature*; AENOR: Madrid, Spain, 2017.
31. ASTM. *D1141-98 Standard Practice for the Preparation of Substitute Ocean Water*; American Society for Testing Materials: West Conshohocken, PA, USA, 2013.
32. ASTM. *ASTM D1193-06(20018) Standard Specification for Reagent Water*; Reapproved; American Society for Testing Materials: West Conshohocken, PA, USA, 2018.
33. ASTM. *G129-00(2013) Standard Practice for Slow Strain Rate Testing to Evaluate the Susceptibility of Metallic Materials to Environmentally Assisted Cracking*; American Society for Testing Materials: West Conshohocken, PA, USA, 2013.

34. Álvarez, J.A.; Lacalle, R.; Arroyo, B.; Cicero, S.; Gutiérrez-Solana, F. Failure analysis of high strength galvanized bolts used in steel towers. *Metals* **2016**, *6*, 163. [CrossRef]

Publisher's Note: MDPI stays neutral with regard to jurisdictional claims in published maps and institutional affiliations.

© 2020 by the authors. Licensee MDPI, Basel, Switzerland. This article is an open access article distributed under the terms and conditions of the Creative Commons Attribution (CC BY) license (http://creativecommons.org/licenses/by/4.0/).

Article

The Evaluation of Front Shapes of Through-the-Thickness Fatigue Cracks

Behnam Zakavi [1,*], Andrei Kotousov [1] and Ricardo Branco [2]

[1] School of Mechanical Engineering, The University of Adelaide, Adelaide, SA 5005, Australia; andrei.kotousov@adelaide.edu.au

[2] Department of Mechanical Engineering, CEMMPRE, The University of Coimbra, 3030-788 Coimbra, Portugal; ricardo.branco@dem.uc.pt

* Correspondence: behnam.zakavi@adelaide.edu.au; Tel.: +61-88-313-5439

Abstract: Fatigue failure of structural components due to cyclic loading is a major concern for engineers. Although metal fatigue is a relatively old subject, current methods for the evaluation of fatigue crack growth and fatigue lifetime have several limitations. In general, these methods largely disregard the actual shape of the crack front by introducing various simplifications, namely shape constraints. Therefore, more research is required to develop new approaches to correctly understand the underlying mechanisms associated with the fatigue crack growth. This paper presents new tools to evaluate the crack front shape of through-the-thickness cracks propagating in plates under quasi-steady-state conditions. A numerical approach incorporating simplified phenomenological models of plasticity-induced crack closure was developed and validated against experimental results. The predicted crack front shapes and crack closure values were, in general, in agreement with those found in the experimental observations.

Keywords: crack front shape; structural plates; through-the-thickness crack; steady-state loading conditions; small-scale yielding

1. Introduction

The evaluation of fatigue life and failure conditions of structural components is of permanent and primary interest for engineers. Over the past five decades, significant progress has been made toward the development of more appropriate fatigue crack growth models and life assessment procedures. Significant research effort has been directed to the study of the fatigue crack closure phenomenon, which was first introduced by Elber [1] to explain the experimentally observed features of fatigue crack growth in aluminium alloys. The number of publications grew rapidly since this pioneering study, and continues to grow. It is now commonly accepted that the contributions of various crack closure mechanisms, specifically plasticity-induced crack closure, roughness-induced crack closure, and oxide-induced closure, are significant, and these mechanisms are capable of explaining many fatigue crack growth phenomena, e.g., the influence of thickness on crack growth rates, retardation effects associated with overloads, or higher propagation rates of small cracks in comparison with long cracks [2].

It is well-established that for relatively long cracks propagating in a non-aggressive environment, the plasticity-induced crack closure dominates over the roughness-induced crack and oxide-induced closures. The plasticity-induced crack closure models rely on far fewer assumptions than the two other closure mechanisms. The first theoretical model was developed by Budianski and Hutchinson [3] based on the two-dimensional Dugdale strip-yield model [4]. The theoretical results demonstrated that opening stress intensity factor is surprisingly high, and increases with an increase in the R ratio. All early crack closure models for plate components utilised both plane strain and plane stress simplifications, although real cracks are inherently three-dimensional (3D). To examine the thickness effect on crack propagation rates, empirical constraint factors were often used, demonstrating a stronger correlation with experimental results. With the advance of numerical methods and the increase in computational power, it became possible to eliminate these simplifications and study more realistic geometries, as well as various 3D effects [5,6].

In 3D problems, the order of the singularity at the intersection of the crack front with the free surface depends on Poisson's ratio and the intersection angle. From energy considerations, it follows that fatigue cracks have to preserve the $1/\sqrt{r}$ singularity. Therefore, the fatigue crack has to intersect the free surface at a critical angle, β_{cr}, which is a function of Poisson's ratio. Several experimental studies have reported that, at least, the Mode I fatigue crack front is shaped to ensure the square root singular behaviour along the entire crack front. However, it seems that the effect of 3D corner singularity is not very significant in the presence of a sufficiently large crack front process zone [7]. This is because the 3D corner singularity effect is a point effect, and is very much localised. Therefore, it might be negated by the plasticity and damage formation near the surface. For example, in an experimental study of steel circular bars subjected to bending and torsion, the experimental intersection angles were found to be very different from the theoretically predicted critical angles [8].

Considerably less effort has been directed toward the study of the effects of the 3D corner singularity and elasto-plastic constraints on plasticity-induced crack closure. Generally, the direct 3D elasto-plastic simulations of fatigue crack growth demand much greater computational resources [9]. These simulations have many issues associated with the validation of the numerical solution and the accuracy of the obtained results. A number of factors affect the accuracy, which are difficult to control: the mesh refinement, the type of finite element, the crack advance scheme (which usually consists of releasing nodes ahead of the crack front), contact conditions, and the local criterion of crack front opening. Branco et al. [10] recently provided an exhaustive review concerning these aspects. The overall conclusion was that the direct numerical approaches are capable of describing the shape evaluation of fatigue cracks. However, the application of these approaches to particular problems can be quite cumbersome. Each problem needs a large effort to calibrate the solution and verify the results. These efforts are usually focused on the reduction in the number of finite elements, the number of simulations required in the analysis, or, eventually, the computation time, which cannot be considered to be of practical relevance [11].

In the present paper, a simplified procedure for the evaluation of the fatigue crack front shapes of through-the-thickness cracks propagating under the cyclic loading conditions is presented. The procedure is based on simplified methods for the evaluation of the plasticity-induced crack closure effect, namely the equivalent-thickness method introduced by Yu and Guo [12,13], as well as the analytical model developed by Kotousov et al. [14,15]. The outcomes of the simulation are compared with available experimental results obtained at the same propagation conditions for validation purposes. The paper is organised as follows: Section 2 addresses the method used to evaluate the crack front shape, as well as the models introduced to evaluate the crack closure along the crack front. Section 3 describes the finite element model developed to calculate the stress intensity factors along the crack front. Section 4 compares the predicted crack front shapes with those obtained experimentally for different materials and propagation conditions. The paper ends with some concluding remarks.

2. Crack Shape Simulation and Crack Closure Models

The main idea behind the evaluation of the steady-state shape of a fatigue crack front proposed in this paper is to select a curve from a parametric family that minimises the deviation of the fatigue driving force along the crack front. In other words, we first specified a possible parametric set of curves in the crack plane (e.g., parabolic, hyperbolic, or elliptical shapes) and then evaluated the local fatigue driving force using a finite element model, along with simplified plasticity-induced crack closure models. In this study, the local fatigue driving force was defined by the effective stress intensity factor range, ΔK_{eff}, given by the formula:

$$\Delta K_{eff} = U \cdot \Delta K = U \cdot (K_{max} - K_{min}) \quad (1)$$

where U is the normalised load ratio parameter, or normalised effective stress intensity factor, which is often used to describe the effects of loading and geometry on crack closure, and ΔK is the traditional linear-elastic stress intensity factor range [16] defined by the maximum and minimum values of the stress intensity factor experienced for a given load cycle. The load ratio is therefore given by $R = K_{min}/K_{max}$. In the case of 3D problems, this normalised load ratio is not a constant, but rather a function of the position along the crack front, $U = U(z)$. Thus, the local crack growth rate is a function of the effective stress intensity factor range, i.e.,

$$\Delta K_{eff}(z) = K_{max}(z) - K_{op}(z) = U(z)\Delta K(z) \quad (2)$$

where $K_{op}(z)$ is the local opening load stress intensity factor, which corresponds to the minimum load at which the crack faces, at point z, which are fully separated.

A number of sophisticated finite-element (FE) models were developed to evaluate $U(z)$ for different geometries and loading conditions. However, as discussed above, these models have many limitations, and are quite difficult to apply in fatigue calculations. Below, we consider two simplified methods for the evaluation of the normalised load ratio, the equivalent-thickness model introduced by Yu and Guo [13], and the analytical model proposed by Kotousov et al. [14,15], which are addressed in Sections 2.1 and 2.2, respectively. These methods will be further incorporated into the 3D linear elastic finite element simulations to evaluate the shape of the through-the-thickness cracks. This evaluation will be performed via the corner singularity method [17], which is briefly presented in Section 2.3.

2.1. Equivalent-Thickness Model

For through-the-thickness cracked plates, She et al. [17] proposed defining the equivalent thickness based on a numerical analysis of the 3D distribution of the out-of-plane stresses and constraint factor, T_z, which is defined as:

$$T_z = \frac{\sigma_z}{\sigma_x + \sigma_y} \quad (3)$$

where σ_x, σ_y, and σ_z are the normal stresses. This method is illustrated in Figure 1. The equivalent thickness, $2h_{eq}$, for point P on the crack front is identified as the plate thickness, which leads to the same distribution of T_z at the mid-plane.

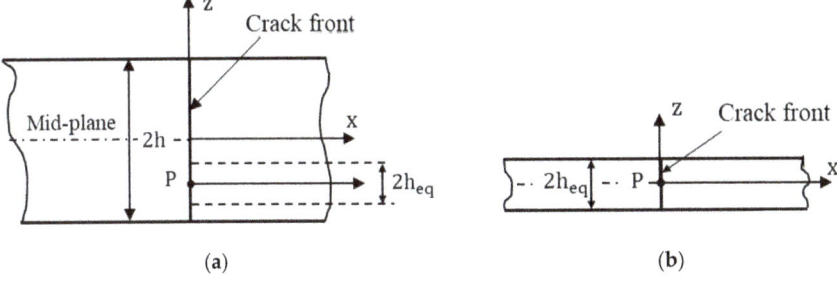

Figure 1. Schematic illustration of the equivalent-thickness method in the through-the-thickness cracks: (**a**) original straight through-the-thickness cracked geometry; (**b**) final straight-through-the-thickness cracked geometry with equivalent thickness.

An empirical equation was suggested to evaluate the equivalent-thickness as follows:

$$\frac{h_{eq}}{h} = 1 - \left(\frac{z}{h}\right)^2 \quad (4)$$

where z is the distance from the mid-plane and h is the half-thickness of the plate. The normalised load ratio parameter in this method can be calculated as follows:

$$U = \frac{\sqrt[3]{\kappa}}{1-R} \quad (5)$$

where κ is a function of the R ratio:

$$\kappa = \frac{(1-R^2)^2(1+10.34R^2)}{\left(1+1.67R^{1.61} + \frac{1}{0.15\pi^2 \alpha_g}\right)^{4.6}} \quad (6)$$

and α_g is a global constraint factor, α_g, defined by the formula:

$$\alpha_g = \frac{1+t}{1-2\nu+t} \quad (7)$$

where ν is the Poisson's ratio and t is given by:

$$t = 0.2088\sqrt{\frac{r_0}{h_{eq}}} + 1.5046\frac{r_0}{h_{eq}} \quad (8)$$

with:

$$r_0 = \frac{\pi}{16}\left(\frac{K_{max}}{\sigma_0}\right)^2 \quad (9)$$

where σ_0 is the flow stress. These empirical equations were extended to the corner, and surface cracks and were extensively validated using 3D finite element analyses.

2.2. Analytical Model for the Evaluation of Crack Closure

Another method for the evaluation of local plasticity-induced closure is based on a simplified 3D analytical model. In accordance with this model, the parameter U for Mode I loading under small-scale yielding conditions can be approximated from the following expression:

$$U(R, \eta) = a(\eta) + b(\eta)R + c(\eta)R^2 \quad (10)$$

where the fitting functions a, b and c can be written in the form:

$$\begin{aligned} a(\eta) &= 0.446 + 0.266 \cdot e^{-0.41\eta} \\ b(\eta) &= 0.373 + 0.354 \cdot e^{-0.235\eta} \\ c(\eta) &= 0.2 - 0.667 \cdot e^{-0.515\eta} \end{aligned} \quad (11)$$

where $\eta = K_{max}/(h\sqrt{\sigma_f})$ is a dimensionless parameter.

The above equations were obtained within the first-order plate theory based on the Budiansky–Hutchinson crack closure model [3,15]. The results, which correspond to the classical two-dimensional theories (or plane stress state, or plane strain state), can be obtained as limiting cases of very thin and very thick plates, i.e., when $\eta \to \infty$ or $\eta \to 0$, respectively. The details of the derivation of these equations can be found in the original paper by Codrington and Kotousov [14].

2.3. Corner Singularity Method

In this study, the evaluation of the steady-state front in the through-the-thickness cracks was carried out using the corner singularity method. First, we approximated the shape of the crack front by a two-parameter elliptical curve, which can be described as:

$$x = b\sqrt{1 - \frac{z^2}{a^2}} \quad -h \leq z \leq h \tag{12}$$

where a and b are the major and minor semi-axes of an ellipse, respectively, as shown in Figure 2.

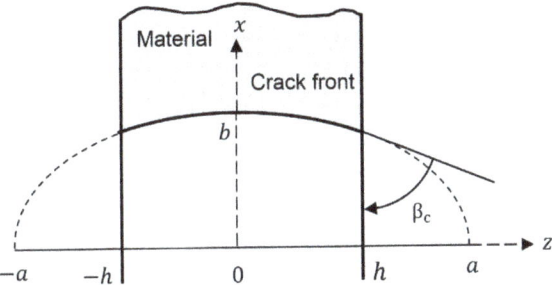

Figure 2. Elliptical-arc crack front shape for geometrical parameters crack propagation.

The crack front tends to intersect the free plate surface at the critical angle, β_c, when the plasticity effects are small. The critical angle is a function of the Poisson's ratio and the type of loading. We found that the critical intersection angle can be approximated by the following formula [18]:

$$\tan \beta_c = \frac{\nu - 2}{\nu} \tag{13}$$

where ν is the Poisson's ratio. Typically, when the size of the plastic zone is greater than 1% of the plate thickness, the stress state near the vertex location is not controlled by the elastic singularity. In these cases, the plasticity effects become more important, and together with the vertex singularity effect, lead to greater critical angles for elastic-plastic materials. To find b, we need to make sure that:

$$\frac{\partial x}{\partial z}\bigg|_{z=\pm h} = -\frac{bh}{a\sqrt{a^2 - h^2}} = \frac{\nu}{\nu - 2} \tag{14}$$

where b is defined by:

$$b = \frac{a\nu}{(2-\nu)}\sqrt{\frac{a^2}{h^2} - 1} \tag{15}$$

Substituting Equation (15) into Equation (12), we obtain:

$$x(z) = \frac{a\nu}{(2-\nu)}\sqrt{\frac{a^2}{h^2} - 1} \cdot \sqrt{a^2 - z^2} \quad -h \leq z \leq h \tag{16}$$

This equation meets the condition that the crack front intersects with the free surface at the critical angle given by Equation (13), and represents a parametric curve with one single parameter, a. Further, the steady-state condition of the crack propagation requires that the projection of the effective stress intensity factor along the crack propagation direction (x-direction, Figure 1) is constant for all points along the crack front. This condition cannot be satisfied exactly with any multi-parametric equation describing the possible crack front shapes. However, the shape that minimises the difference of the effective stress intensity factor along the crack front can be considered as the best approximation of the actual fatigue crack front shape.

3. Numerical Approach

This section describes the numerical model developed in this research to determine the stress intensity factor ranges along the crack front. The stress intensity factor ranges, along with the crack closure models described in the previous section, enabled the computation of the local fatigue driving force, which was used to obtain a steady-state crack front shape. The steady-state crack front shape was selected as the one producing the minimum deviation of ΔK_{eff} along the crack front. This evaluation needs to be completed for each curve from the parametric set.

To reduce the computational overhead, we developed a simplified geometry by introducing adequate boundary conditions, capable of describing 3D effects near the crack front. Section 3.1 describes the details of the numerical modelling, and Section 3.2 addresses the boundary conditions considered in this paper. The last section, Section 3.3, is devoted to the validation of the stress intensity factor values obtained with the proposed approach.

3.1. Finite Element Model Description

The typical finite element geometry, developed here to study a through-the-thickness crack in an elastic plate, is shown in Figure 3. As can be seen, the rectangular cross-section geometries were modelled to evaluate the stress and displacement fields near the crack tip. The size of the finite element models is sufficient to avoid the effect of the finite boundaries on the stress state. By taking advantage of the symmetry conditions (i.e., XY symmetry, XZ symmetry, and YZ symmetry), only one-eighth of the crack problem was modelled. The height of the FE models taken was approximately ten times larger than the plate thicknesses. In accordance with the previous studies, this is sufficient to accurately describe the 3D effects near the crack front [19,20].

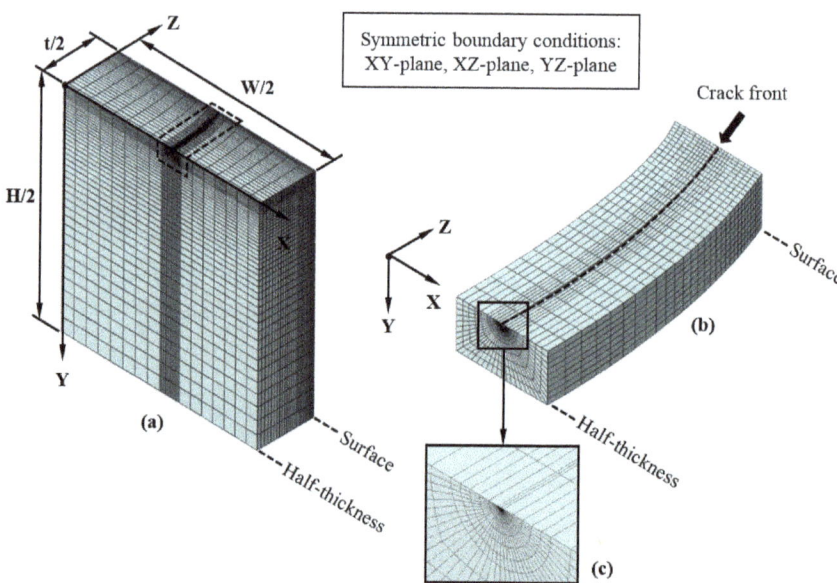

Figure 3. Finite element mesh: (**a**) assembled model; (**b**) detail of the crack front; (**c**) detail of the spider web pattern.

The FE models corresponding to different values of a (Figure 1) were meshed with linear 8-node hexahedral elements of type C3D8R. A reasonably uniform element grid with a structured mesh was considered. A denser mesh, with a spider-web pattern (Figure 3c) was used near the crack front, where the stress gradients were expected to be maximum (Figure 3b), consisting of 5 concentric rings centred at the crack tip with a radial discretisation of 10° (Figure 3c). Thirty nodes along the plate half-thickness (Figure 3b) were used to define the crack front shape. The specimen was subjected to uniaxial loading applied at the bottom surface (i.e., at the XZ-plane with a Y-coordinate equal to $H/2$). The assembled mode is exhibited in Figure 3a. Further details about the modelling approach can be found in papers published by the present authors [19,21].

The numerical simulations were carried out using Abaqus/CAE 2020 (© Dassault Systèmes, 2019), assuming a homogeneous, isotropic, and linear-elastic behaviour. The mechanical properties inserted into Abaqus/CAE 2020 to perform the numerical simulations were the Young's modulus and the Poisson's ratio of the tested materials (Table 1). The displacement field far from the crack tip was calculated in accordance with the William's solution using MATLAB R2020b, and the obtained results were applied for the boundary conditions. The 3D solutions of the J-integral were used to calculate the stress intensity factor near the crack front. One layer of elements surrounding the crack front was used to calculate the first contour integral. The additional layer of elements was used to compute the subsequent contours. The different contour solutions were approximately coincident after eight contours. The results from averaging contours five through eight was considered. A similar strategy, either in terms of mesh framework or simulation analysis, was carried out for all geometries and crack configurations studied in the present paper.

Table 1. Mechanical properties of the selected materials.

Material	Young's Modulus, E	Poisson's Ratio, ν	Fracture Toughness, K_{IC}	Exponent of Paris Law, m
6082-T6	74 GPa	0.33	20 MPa·m$^{0.5}$	3.456
PMMA	3.6 GPa	0.365	1.6 Pa·m$^{0.5}$	0.91

3.2. Boundary Conditions

The plane-stress displacements far from the crack tip were calculated in accordance with William's solution [22]:

$$u_x(r,\theta) = \left(\frac{r}{2\pi}\right)^{1/2} \frac{(1+\vartheta)}{E}[K_I^\infty f_x^I(\theta)] \tag{17}$$

$$u_y(r,\theta) = \left(\frac{r}{2\pi}\right)^{1/2} \frac{(1+\nu)}{E}[K_I^\infty f_y^I(\theta)] \tag{18}$$

Being:

$$f_x^I(\theta) = \cos\frac{\theta}{2}\left(k - 1 + 2\sin^2\frac{\theta}{2}\right) \tag{19}$$

$$f_y^I(\theta) = \sin\frac{\theta}{2}\left(k + 1 + 2\cos^2\frac{\theta}{2}\right) \tag{20}$$

where r is the distance from the crack tip, θ is the angle measured from the symmetry line, K_I^∞ is the remotely applied Mode I stress intensity factor, and k is Kolosov's constant for plane stress and plane strain conditions. The plane stress k value was considered in the boundary conditions, i.e.,

$$k = \frac{3-\nu}{1+\nu} \tag{21}$$

where ν is the Poisson's ratio. Bakker [23] showed that a cracked plate under plane stress undergoes a change to plane strain behaviour near the crack tip. He proved that the radial position, where the plane stress to plane strain transition takes place, strongly depends on the position in the thickness direction. The degree of plane strain is essentially zero at distances from the tip greater than five times the thickness, even in the middle plane of the plate [24].

3.3. Validation Study

The numerical results obtained for the maximum stress intensity factor are presented in Figure 4 as a function of the thickness for a Poisson's ratio of 0.3. The classical results for both the plane stress state and the plane strain state are also given in Figure 4. It is evident from Figure 4 that the stress intensity factor changes with the thickness of the plate until the thickness exceeds a critical value. In this particular problem, the results showed that the critical thickness is 25 mm. Once the thickness exceeds the critical dimension, the stress field in the vertex singularity region has a negligible impact on the behavior of the whole structure. The stress intensity factor becomes relatively constant in the sufficiently thick plate, and is equal to the value for plane strain conditions.

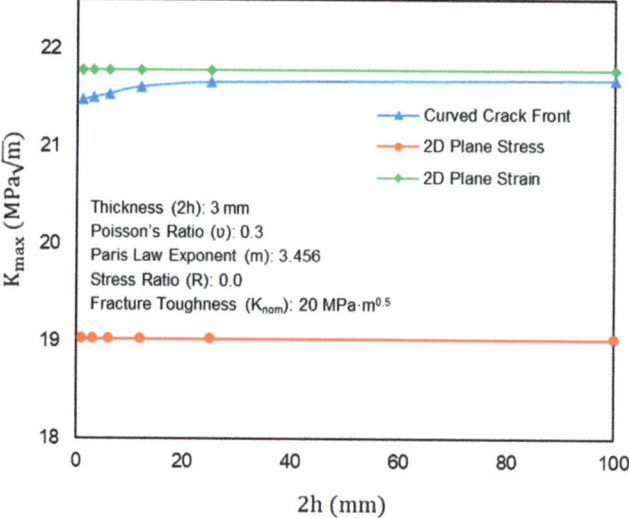

Figure 4. The effect of the thickness on the maximum stress intensity factor.

4. Crack Front Shape Evaluation and Comparison with Experimental Studies

The proposed method for the evaluation of the steady-state crack front shapes was compared against two independent experimental studies. The specimen geometries used in the experimental tests are exhibited in Figure 5, and were made of 6082-T6 aluminium alloy and polymethyl methacrylate (PMMA), separately. The main mechanical properties of both materials are listed in Table 1. The former (Figure 5a) consisted of a standard middle-crack tension specimen with a thickness of 3 mm [11,25]. The tests were conducted under constant-amplitude axial loading using a stress ratio equal to 0.25. Figure 6a shows an example of the typical fracture surfaces obtained in the tests. Fatigue cracks grew over a sufficiently large distance from the initial notch to ensure the quasi-steady-state conditions of propagation. The beach-marking technique was applied to mark the crack front at the fracture surface.

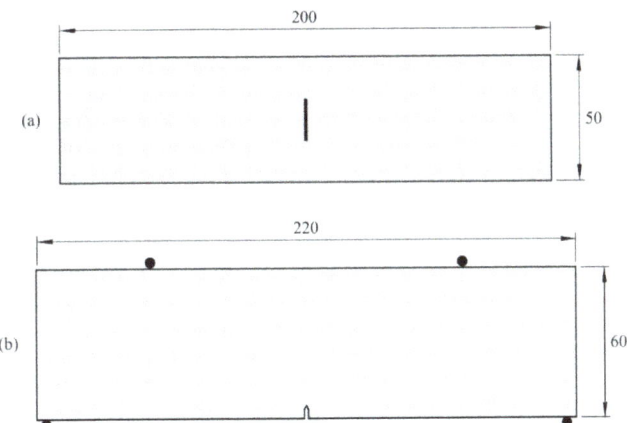

Figure 5. Specimen geometries used in the crack front shape evaluation: (**a**) 6082-T6 aluminium alloy and (**b**) polymethyl methacrylate (PMMA). All dimensions are in mm.

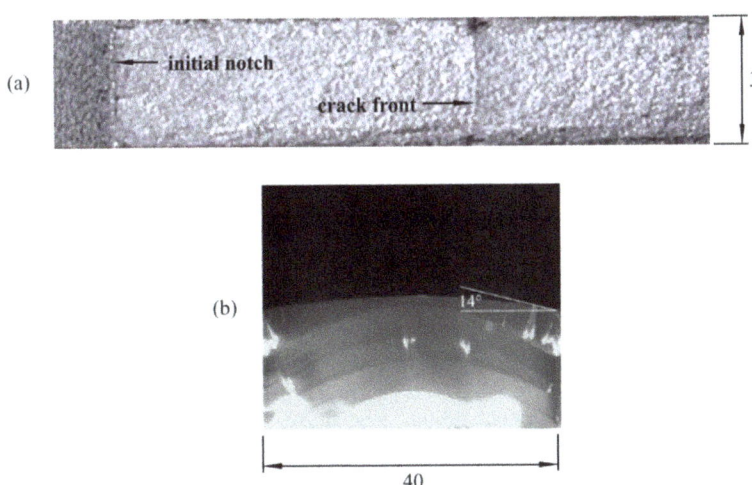

Figure 6. The crack front shapes observed in the experiments for the: (**a**) 6082-T6 aluminium alloy reprinted with permission from ref. [11], copyright 2021 Elsevier and (**b**) polymethyl methacrylate reprinted with permission from ref. [26], copyright 2021 Elsevier. Propagation direction is from left to right in case (**a**) and from bottom to top in case (**b**). All dimensions are in millimetres.

Regarding the latter (Figure 5b), the specimen geometry was made of polymethyl methacrylate. It had a rectangular cross-section (Figure 5b), with a thickness of 40 mm [26,27], and an initial straight notch at the middle of the specimen. The tests were conducted under four-point bending loading conditions using a stress ratio equal to 0. The crack front shape was evaluated in situ using a high-resolution digital camera. As in the previous case, fatigue cracks propagated over a sufficiently large distance from the initial notch to ensure the quasi-steady state conditions of propagation. An example of the crack front shapes observed in the experiments is exhibited in Figure 6b.

Figure 7a,b displays a comparison of the experimental crack front shapes and those obtained with the proposed methods for the 6082-T6 aluminium alloy and PMMA, respectively. Overall, the results showed that the equivalent-thickness method provides a satisfactory approximation for the fatigue crack propagation under small yielding conditions. Moreover, the experimental results confirmed that the angle at which the crack front intersects the free surface is greater than the proposed empirical equations in the sufficiently plastic materials. We think that the careful combination of the hyperbolic and elliptical functions might provide accurate crack front shape estimation in the presence of residual stresses or large crack closure effects. The good agreement demonstrated in the previous analysis confirmed the possibility of the accurate evaluation of stress intensity factors using the proposed approach in materials controlled by 3D corner singularity effects.

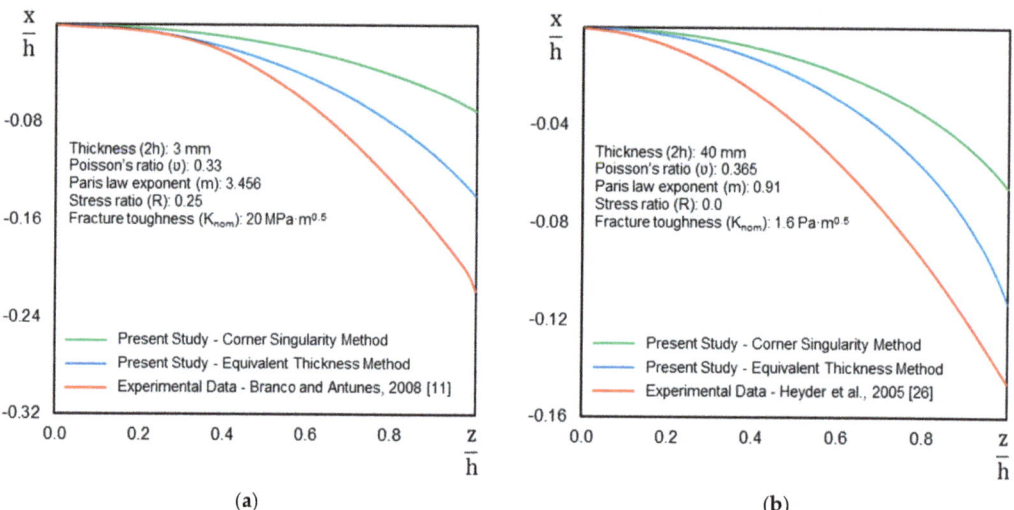

Figure 7. A comparison between the predicated crack shapes and experimental data for the specimens composed of: (**a**) 6082-T6 aluminium alloy and (**b**) polymethyl methacrylate.

This methodology can also be applied to conduct parametric studies associated with the main variables affecting the fatigue crack growth of through-the-thickness cracks. A subject that can be analysed with the developed approach is the effect of the stress ratio on crack closure values. Figure 8 plots the ratio of the opening stress intensity factor (K_o) to the maximum stress intensity factor (K_{max}) along the crack front for both materials. As shown, the plane stress curve represents the upper limit, while the plane strain curve represents the lower limit. The values of K_o/K_{max} are between two limiting cases, and decrease with an increase in the stress ratio. In addition, at lower stress ratios, the differences between the maximum and minimum values of K_o/K_{max} are higher for PMMA and tend to be closer for the aluminium alloy.

Figure 9 plots the variation in the K_o/K_{max} ratio at the crack surface obtained from the presented 3D FE simulations against previously published relationships based on experimental tests that incorporated plasticity-induced crack closure. Notably, the results of the presented procedure agree well with the outcomes of the experimental and theoretical studies reported in the literature [1,16,27–29]. The variation between the presented method and published data decreases with an increase in the R ratio, as the size of the reverse plasticity zone (or monotonic plastic zone) becomes smaller in the fatigue crack growth rates. These results provide further support to and validation of the numerical technique outlined in this paper.

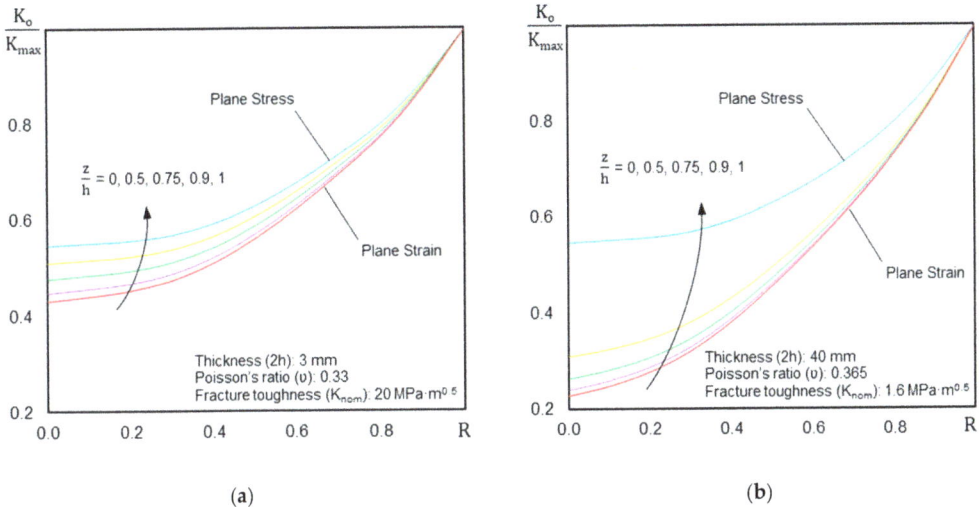

Figure 8. The ratio of the opening stress intensity factor to the maximum stress intensity factor as a function of the R ratio along the crack front: (**a**) 6082-T6 aluminium alloy; (**b**) polymethyl methacrylate.

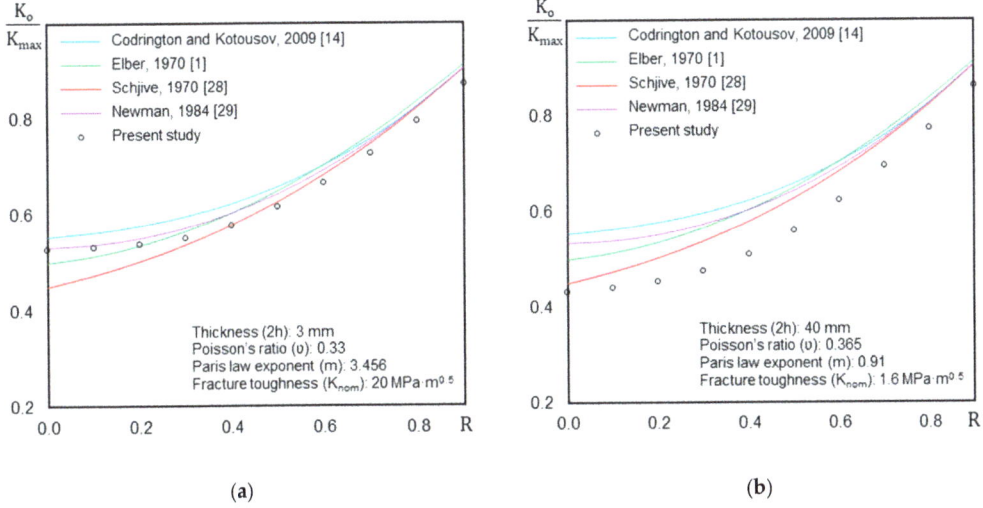

Figure 9. The ratio of the of the opening stress intensity factor to the maximum stress intensity factor as a function of the R ratio along the crack front and past published functions: (**a**) 6082-T6 aluminium alloy; (**b**) polymethyl methacrylate.

5. Conclusions

In this paper, new numerical modelling tools capable of simulating the crack shape development of through-the-thickness fatigue cracks in finite plates were presented. The proposed approaches assume a pre-defined crack front shape, and include plasticity-induced crack closure. The methodology was successfully tested for cracked rectangular cross-section geometries when subjected to Mode I loading. The following conclusions can be drawn:

1. The maximum stress intensity factor becomes relatively constant in the sufficiently thick plates and is equal to the value obtained for plane strain state conditions. The

plane strain fatigue models (2D) may lead to inaccurate predictions when applied to the analysis of fatigue crack growth of thin structural plates;
2. The proposed methodology leads to satisfactory crack front predictions, either for ductile materials or brittle materials. Moreover, it is sensitive to the plate thickness, enabling good results for both thin and thicker geometries. In addition, it is capable of dealing with different stress ratios;
3. The opening stress intensity factor increases with increasing values of stress ratio, maximum stress intensity factor, and distance from the centre of the crack. Predicted values obtained by the proposed methodology are quite close to those found in the literature for the same propagation conditions.

The comparison with experimental results is encouraging, and demonstrates the validity of the underlying assumptions: (1) the crack front shape intersects the free plate surface at the critical angle; ad (2) the local stress intensity factor can be considered as the fatigue crack driving force, which leads to the formation of the crack front shape under high cycling loading. The above assumptions might not be correct in the case of large plastic effects near the crack tip. In this case, the plasticity-induced crack closure, which is significantly different along the crack front, will be the one of the most influential factors affecting the crack front shape.

Future work will be directed to the application of the proposed methodology to more complex problems in terms of geometry, loading scenario, and crack shape configuration. Lastly, the simplicity and speed of calculation of the proposed approach, compared to the current numerical solutions used for the same purpose, make it quite attractive for simulating the fatigue crack growth, in both practical applications and parametric studies.

Author Contributions: Conceptualization, B.Z. and A.K.; methodology, B.Z. and A.K.; software, B.Z; validation, B.Z. and R.B.; formal analysis, B.Z. and R.B.; investigation, B.Z. and R.B.; data curation, B.Z. and R.B.; writing—original draft preparation, B.Z.; writing—review and editing, B.Z., R.B. and A.K.; visualization, B.Z., A.K. and R.B.; supervision, A.K. All authors have read and agreed to the published version of the manuscript.

Funding: This research was sponsored by FEDER funds through the program COMPETE (Programa Operacional Factores de Competitividade) and by national funds through FCT (Fundação para a Ciência e a Tecnologia) under the project UIDB/00285/202.

Data Availability Statement: The data presented in this study are available from the corresponding author, upon reasonable request. The data are not publicly available due to ethical restrictions.

Conflicts of Interest: The authors declare no conflict of interest.

References

1. Elber, W. Fatigue crack closure under cyclic tension. *Eng. Fract. Mech.* **1970**, *2*, 37–44. [CrossRef]
2. Vasudeven, A.K.; Sadananda, K.; Louat, N. A review of crack closure, fatigue crack threshold and related phenomena. *Mater. Sci. Eng. A* **1994**, *188*, 1–22. [CrossRef]
3. Budiansky, B.; Hutchinson, J.W. Analysis of closure in fatigue crack growth. *ASME J. Appl. Mech.* **1978**, *45*, 267–276. [CrossRef]
4. Dugdale, D.S. Yielding of steel sheets containing slits. *J. Mech. Phys. Solids* **1960**, *8*, 100–104. [CrossRef]
5. He, Z.; Kotousov, A.; Berto, F.; Branco, R. A brief review of three-dimensional effects near crack front. *Phys. Mesomech.* **2019**, *19*, 6–20. [CrossRef]
6. Kotousov, A.; Khanna, A.; Branco, R.; Jesus, A.; Correia, J.A. Review of current progress in 3D linear elastic fracture mechanics. In *Mechanical Fatigue of Metals, Structural Integrity*; Springer: Cham, Switzerland, 2019; pp. 125–131. [CrossRef]
7. Nowell, D.; de Matos, P.F.P. The influence of the Poisson's ratio and corner point singularities in three-dimensional plasticity-induced fatigue crack closure: A numerical study. *Int. J. Fatigue* **2008**, *30*, 1930–1943. [CrossRef]
8. Lebahn, J.; Heyer, H.; Sander, M. Numerical stress intensity factor calculation in flawed round bars validated by crack propagation tests. *Eng. Fract. Mech.* **2013**, *108*, 37. [CrossRef]
9. Camas, D.; Garcia-Manrique, J.; Antunes, F.V.; Gonzalez-Herrera, A. Three-dimensional fatigue crack closure numerical modelling: Crack growth scheme. *Theor. Appl. Fract. Mech.* **2020**, *108*, 102623. [CrossRef]
10. Branco, R.; Antunes, F.V.; Costa, J.D. A review on 3D-FE adaptive remeshing techniques for crack growth modelling. *Eng. Fract. Mech.* **2015**, *141*, 170–195. [CrossRef]

11. Branco, R.; Antunes, F.V. Finite element modelling and analysis of crack shape evolution in Mode-I fatigue Middle Cracked Tension specimens. *Eng. Fract. Mech.* **2008**, *75*, 3020–3037. [CrossRef]
12. Guo, W. Three-dimensional analyses of plastic constraint for through-thickness cracked bodies. *Eng. Fract. Mech.* **1999**, *62*, 383–407. [CrossRef]
13. Yu, P.; Guo, W. An equivalent thickness conception for evaluation of corner and surface fatigue crack closure. *Eng. Fract. Mech.* **2013**, *99*, 202. [CrossRef]
14. Codrington, J.; Kotousov, A. A crack closure model of fatigue crack growth in plates of finite thickness under small-scale yielding conditions. *Mech. Mater.* **2009**, *41*, 165–173. [CrossRef]
15. He, Z.; Kotousov, A.; Branco, R. A simplified method for the evaluation of fatigue crack front shapes under mode I loading. *Int. J. Fract.* **2014**, *188*, 203–211. [CrossRef]
16. Antunes, F.V.; Chegini, A.G.; Correia, L.; Branco, R. Numerical study of contact forces for crack closure analysis. *Int. J. Solids Struct.* **2014**, *51*, 1330–1339. [CrossRef]
17. She, C.; Zhao, J.; Guo, W. Three-dimensional stress fields near notches and cracks. *Int. J. Fract.* **2008**, *151*, 151–160. [CrossRef]
18. Pook, L.P. Some implications of corner point singularities. *Eng. Fract. Mech.* **1994**, *48*, 367–378. [CrossRef]
19. Kotousov, A.; Lazzarin, P.; Berto, F.; Pook, L.P. Three-dimensional stress states at crack tip induced by shear and anti-plane loading. *Eng. Fract. Mech.* **2013**, *108*, 65–74. [CrossRef]
20. Nakamura, T.; Parks, D.M. Antisymmetrical 3-D stress field near the crack front of a thin elastic plate. *Int. J. Solids Struct.* **1989**, *25*, 1411–1426. [CrossRef]
21. Kotousov, A. Fracture in plates of finite thickness. *Int. J. Solids Struct.* **2007**, *44*, 8259–8273. [CrossRef]
22. Williams, M.L. On the stress distribution at the base of a stationary crack. *J. Appl. Mech.* **1957**, *24*, 109–114. [CrossRef]
23. Bakker, A. Three dimensional constraint effects on stress intensity distributions in plate geometries with through-thickness cracks. *Fatigue Fract. Eng. Mater. Struct.* **1992**, *15*, 1051–1069. [CrossRef]
24. Levy, N.; Marcal, P.V.; Rice, J.R. Progress in three-dimensional elastic-plastic stress analysis for fracture mechanics. *Nucl. Eng. Des.* **1971**, *17*, 64–75. [CrossRef]
25. Borrego, L.F. Fatigue Crack Growth under Variable Amplitude Load in an AlMgSi Alloy. Ph.D. Thesis, University of Coimbra, Coimbra, Portugal, 2001.
26. Heyder, M.; Kolk, K.; Kuhn, G. Numerical and experimental investigations of the influence of corner singularities on 3D fatigue crack propagation. *Eng. Fract. Mech.* **2005**, *72*, 2095–2105. [CrossRef]
27. Heyder, M.; Kuhn, G. 3D fatigue crack propagation: Experimental studies. *Int. J. Fatigue* **2006**, *28*, 627–634. [CrossRef]
28. Schjive, J. Some formulas for the crack opening stress level. *Eng. Fract. Mech.* **1981**, *14*, 461–465. [CrossRef]
29. Newman, J.C. A crack opening stress equation for fatigue crack growth. *Int. J. Fatigue* **1984**, *24*, R131–R135. [CrossRef]

Article

Analysis of the Deceleration Methods of Fatigue Crack Growth Rates under Mode I Loading Type in Pearlitic Rail Steel

Grzegorz Lesiuk [1,*], Hryhoriy Nykyforchyn [2], Olha Zvirko [2], Rafał Mech [1], Bartosz Babiarczuk [1], Szymon Duda [1], Joao Maria De Arrabida Farelo [1,3] and Jose A.F.O. Correia [4]

1. Faculty of Mechanical Engineering, Wroclaw University of Science and Technology, PL50370 Wroclaw, Poland; Rafał.Mech@pwr.edu.pl (R.M.); Bartosz.Babiarczuk@pwr.edu.pl (B.B.); Szymon.Duda@pwr.edu.pl (S.D.); 256609@student.pwr.edu.pl (J.M.D.A.F.)
2. Karpenko Physico-Mechanical Institute of the National Academy of Sciences of Ukraine, UA79060 Lviv, Ukraine; nykyfor@ipm.lviv.ua (H.N.); olha.zvirko@gmail.com (O.Z.)
3. Instituto Politécnico de Setúbal-Polytechnic Institute of Setúbal, PT 2910761 Setúbal, Portugal
4. Faculty of Engineering, University of Porto, PT 4200465 Porto, Portugal; jacorreia@fe.up.pt
* Correspondence: grzegorz.lesiuk@pwr.edu.pl; Tel.: +48-71-320-3919

Abstract: The paper presents a comparison of the results of the fatigue crack growth rate for raw rail steel, steel reinforced with composite material—CFRP—and also in the case of counteracting crack growth using the stop-hole technique, as well as with an application of an "anti-crack growth fluid". All specimens were tested using constant load amplitude methods with a maximum loading of F_{max} = 8 kN and stress ratio R = $\sigma_{min}/\sigma_{max}$ = 0.1 in order to analyze the efficiency of different strategies of fatigue crack growth rate deceleration. It has been shown that the fatigue crack grows fastest in the case of the raw material and slowest in the case of "anti-crack growth fluid" application. Additionally, the study on fatigue fracture surfaces using light and scanning electron (SEM) microscopy to analyze the crack growth mechanism was carried out. As a result of fluid activity, the fatigue crack closure occurred and significantly decreased crack driving force and finally resulted in fatigue crack growth decrease.

Keywords: pearlitic steel; CFRP patches; crack retardation; fatigue crack growth; failure analysis

1. Introduction

It is commonly known that each industry, e.g., automotive, marine, building, rail, etc., requests specific parameters and types of different materials, including steel. There are many examples of product orientation in terms of loads, strength, or wear resistance, e.g., in the railway industry. It is expected to have high strength and high wear resistance (rails and wheels of vehicles moving on them). However, much higher focus will be placed on welding properties and corrosion resistance in the marine industry. Pearlitic steel is one of the standard steels which is usually used in places where high loads and wear are expected. It is feasible for this material to work in conditions where it might be exposed to high loads, mainly where high loads occur in a small area. These types of loads cause a local accumulation of stresses, which often exceed the yield point's value. Unfortunately, preventing plastic deformation is not easy or, in many cases, it is simply impossible to avoid. In connection with this, more and more research is carried out on the effects of these deformations, namely cracks and their propagation. The ability to predict the places of crack formation and their propagation paths allows for a sufficiently quick reaction before a tragic catastrophe occurs (break or simply the destruction of the element). The development of fracture mechanics and methods of predicting and determining the lifetime of components under the influence of fatigue undoubtedly contributes to the improvement of safety and reduction of operating costs [1].

Safety, reduction of costs, and changing technologies require optimizing different types of materials in each industry. To optimize and find the proper material for a specific application, many different tests are required. There are many studies on the wear rate for different types of steel. These tests are conducted in laboratory conditions [2,3] and, importantly, in real conditions [4,5]. Comparing the results achieved in laboratories with real conditions greatly helps in achieving the complete characteristics of the tested materials. Other studies mainly focus on fracture toughness [6–8] or focus on fatigue crack growth (FCG) [9]. For example, in [7] it was shown that the fracture toughness of very good quality pearlitic steels is at the level of 30.4 MPa$\sqrt{m}$ at the temperature of −20 °C.

Additionally, another study, [6], investigated how the high shear deformation affects the fracture toughness Kq. The tests were carried out for R260 steel, and the obtained Kq value of the undeformed material varied from 53 to 42 MPa$\sqrt{m}$. These values were obtained after one pass in flush-channel angular pressing (ECAP). It should be noted that for these tests, the samples used were relatively thin (B = 2 mm). Therefore, unfortunately, it is not possible to determine the actual K_{Ic}. Moreover, it is worth noting that the degree of deformation and orientation are of great importance for fracture toughness, which was shown in [8].

The paper presents a comparison of the results of the fatigue crack growth rate for R260 steel, steel reinforced with composite material, and also in the case of counteracting crack growth using the stop-hole technique. Generally speaking, all methods are widely used as deceleration methods of fatigue crack growth. Carbon fiber-reinforced polymer (CFRP) patches are still of particular interest in various applications in civil engineering [9–12]. However, there are not many papers devoted to degradation problems in a real operational environment [13–15]. Thus, in some cases, the historic stop-hole technique is still attractive based on simple crack "blunting" and re-initiation period [16–18].

On the other hand, the fatigue crack growth process is strongly associated with the crack driving force and its local condition. In [19–21], authors explained crack growth rate decreasing with an effective stress intensity factor (crack closure concept [22]) based on ΔK_{eff} concept. This assumption allows us to explain, simply using a closure parameter, the role of crack closure in analytical formulas. Recently, the concept of crack closure triggering due to the application of fluid [23] into the crack was successfully validated for low-carbon steel [19]. Therefore, this paper's main goal is to compare several different (physically) approaches in crack growth rate deceleration in pearlitic steel, which might be used for rail manufacturing [24].

Additionally, for selected materials, visual and microscopic inspection of the fracture surfaces were carried out to characterize them and understand how the fracture development and degradation occurred.

2. Materials and Methods

2.1. Materials

The material from which all the samples presented in the study were prepared was taken from the rail that had previously been withdrawn from use. The rail was delivered directly from the manufacturer in a fully operational condition, and its profile conformed to UIC60 [24]. Samples for examination were cut out under cooling with liquid in order to avoid microstructural changes. The chemical composition of the material was spectrally analyzed. The results are summarized in Table 1. Additionally, Table 2 lists the basic static mechanical properties of the tested R260 steel.

Table 1. Chemical composition of the investigated R260 steel.

	Chemical Composition (in % by Weight)							
Element	C	Mn	Si	P	S	Cr	Ni	Mo	Fe
rail steel	0.721	0.873	0.256	0.012	0.005	0.053	0.032	0.011	bal.

Table 2. Mechanical properties of the investigated steel.

Mechanical Properties of the Rail Steel Samples				
property	Tensile strength (MPa)	Yield strength (MPa)	Elongation (%)	Hardness (HV)
rail steel	998	481	14.5	258

The microstructure of the tested steel is shown in Figure 1. In addition to the samples made of the base material, samples were also made that were reinforced with a carbon fiber material. CFRP polymer is an extremely strong and light fiber-reinforced plastic that contains fibers. In general, CFRP composites use thermosetting resins such as epoxy, polyester, or vinyl ester [10]. The Sika® CarboDur® S1014/180 bands were applied to reinforce steel specimens, pultruded carbon fiber plates on an epoxy matrix for structural strengthening of structures. CFRP is highly used in aeronautics because of its weight (much lighter than steel and aluminum) and strength. Carbon fiber is up to five times stronger than typical steel material, however only in specific conditions (along the direction of fibers). The specimens (Figure 2) were reinforced with CFRP considering how the samples were subjected to the cyclic loading and the direction of fibers in composite material. Figure 3 shows steel specimen (base) and Figure 4 reinforced with CFRP. Two types of reinforced specimens, with wider and narrower CFRP strips, were prepared.

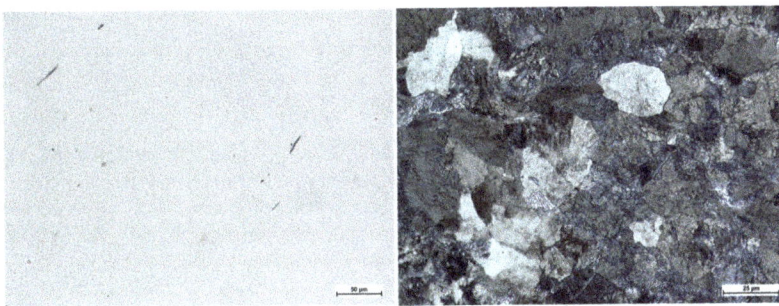

Figure 1. Microstructure of tested steel; non-etched state (on the **left**) with noticeable small non-metallic inclusions and pearlite structure (on the **right**) after etching 3% HNO_3.

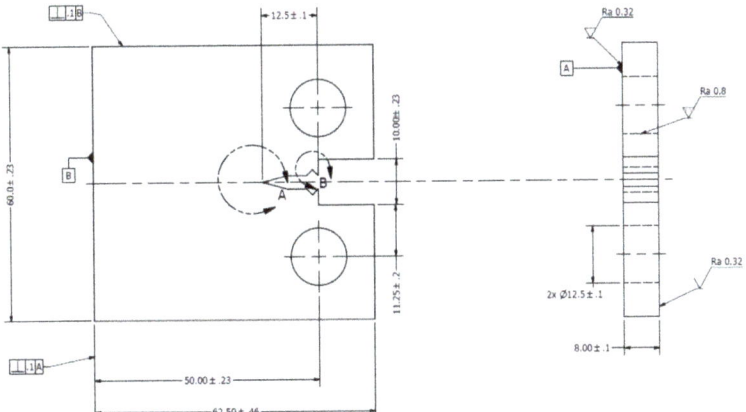

Figure 2. Design of compact tension specimen for fatigue crack growth rate experiment (all dimensions in mm).

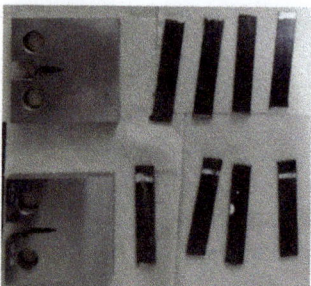

Figure 3. Specimens ready to be glued with carbon fiber-reinforced polymer (CFRP) patches (strips).

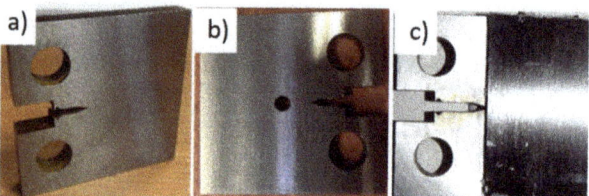

Figure 4. Real specimens devoted to the experimental campaign; (**a**) base metal compact tension (CT), (**b**) specimen after crack growth test—with drilled hole—stope hole technique, (**c**) full-face one-side CFRP patch.

The Sikadur®-31 CF Normal glue was used to connect the CFRP to the steel specimen. This adhesive is a 2-component thixotropic epoxy adhesive. This product is moisture tolerant and based on a combination of epoxy resins and extra filler. The bonding can be utilized in temperatures between +10 °C and +30 °C. It can be applied to join concrete elements, natural stone, ceramics, bricks, mortar, masonry, steel, iron, aluminum, wood, epoxy, and glass. The adhesive has the following advantages:

- easy to apply,
- good adhesion,
- high strength,
- hardens without shrinkage,
- no primer needed,
- good mechanical resistance.

2.2. Methodology

The tests were carried out on five types of samples, which are listed below:

- sample from raw steel material,
- sample from raw steel material with use of the stop-hole technique,
- sample from raw steel material with a wide CFRP patch (full face),
- sample from a raw steel material with a narrow CFRP patch (strip),
- sample from raw steel material with an application of an "anti-crack growth fluid" [23].

For the experimental campaign, compact tension (CT) specimens were prepared in accordance with ASTM E647 [25] standard. The scheme of the specimen is shown in Figure 2. Figures 3 and 4 show prepared specimens ready for tests.

Fatigue tests were carried out on the MTS testing machine with 100 kN capacity. During all tests, data were acquired using MTS TestSuite™ Multipurpose Software (Series 793, MTS Systems, Corporation, Eden Prairie, MN, USA). The straight-through notches of 2.5 mm width and 12.5 mm length were machined and pre-cracked. The pre-crack frequency and maximum stress intensity range were 10 Hz and 15 MPa$\sqrt{m}$, respectively.

For specimens without CFRP stress intensity factor was calculated based on linear elastic fracture mechanics [25]:

$$\Delta K = \frac{\Delta P}{B\sqrt{W}} \frac{(2+\alpha)}{(1-\alpha)^{1.5}} \left[0.886 + 4.64\alpha - 13.32\alpha^2 + 14.72\alpha^3 - 5.6\alpha^4 \right], \quad (1)$$

where α represents the normalized crack length a/W with a as the corresponding crack length observed during the test; ΔP is the applied range of force; B is the thickness of the specimen, and W is the width of specimen defined as in ASTM E647 standard [25]. Crack length was calculated using elastic compliance [25] method automated with MTS system special software for data analysis—Figure 5. Additionally, for specimens with CFRP, crack length was observed in one side (without patch) of the specimen (this strategy was also successfully applied in previous studies based on the beach marking technique in [26]) in order to confirm registered values of crack length.

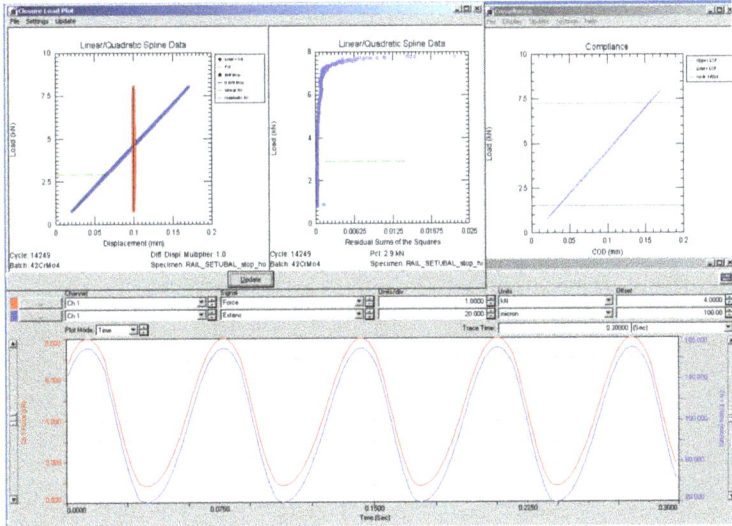

Figure 5. Sinusoidal waveform of loading and measured hysteresis loops and elastic compliance for crack length calculations.

The crack length was monitored using MTS FCGR modular software for hysteresis loop analysis and calculating the current crack length using the compliance method. The exemplary graphical user interface is shown in Figure 5.

This experimental campaign began with standard steel specimens preparation and started with fatigue crack growth rate testing with a maximum stress intensity factor K_{max} = 15 MPa*m$^{0.5}$ (sinusoidal waveform, where R = 0.1) in pre-cracking phase. All types of specimens were pre-cracked up to 14 mm crack length. Then specimens were selected and divided into two groups; one for composite strips (gluing) and another one, "pure metal" and "stop hole" were devoted for direct testing up to a specific number of 75,000 cycles. In this FCGR (fatigue crack growth rate) experiment phase, a constant amplitude loading method was used. F_{max} = 8 kN and R = 0.1 were kept during the experiment for all types of specimens in order to maintain the same testing conditions for all types of specimens.

After reaching mentioned 75,000 cycles in specimen marked as "stop hole" a hole of 5.6 mm diameter was drilled. After drilling the hole, the specimen was placed again on the testing machine, and cyclic loading was continued in order to receive information about fatigue crack growth within the use of the stop-hole technique.

Specimen marked as "with_fluid" (Figures 6 and 7) was tested in special environmental liquid solutions. Unique, patented fluid [23] was tested in order to trigger the artificial crack closure phenomenon. The liquid matter, as the particular technological environment (STE), containing an active component patented in [23] was used in the experiment. The solvent serves as a transport medium of the active component into fatigue crack. After applying the fluid into a crack cavity, it starts to interact with the crack's metal surfaces chemically.

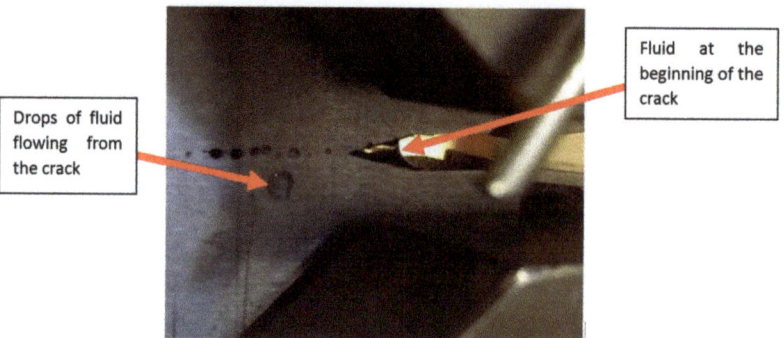

Figure 6. Specimen with fluid injected into the machined notch tip of CT specimen mounted in the hydraulic pulsator.

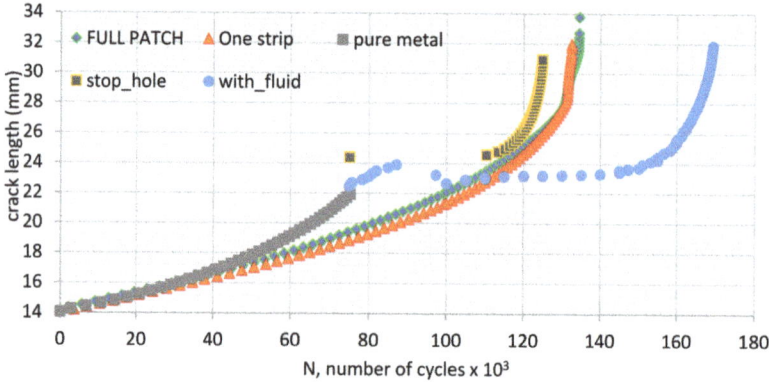

Figure 7. Fatigue crack growth curves for all tested specimens.

In the case of these investigations, the sample did not require any additional operations prior to testing, besides well finishing of sides surfaces of the specimen. The sample was subjected to the same cycling loading values as in the previously mentioned methods. After a specified number of cycles (70,000 cycles), the injections were stopped, and the sample was cycled until the break.

3. Fatigue Crack Growth Results and Fractography

Figure 7 presents fatigue lifetime curves for all specimens tested in the same loading conditions. As it is noticeable comparable effect is obtained for the designed stop hole technique and composite CFRP patches. Noticeable is no extra difference in CFRPs (large and small patch). A similar effect was also observed and described in the Authors' paper [25] with a numerical analysis of the CFRP effect on reducing SIF.

For direct comparison of the fluid activation effect on the crack deceleration for metallic specimens (without CFRP) kinetic fatigue fracture diagram (KFFD) was constructed—Figure 8.

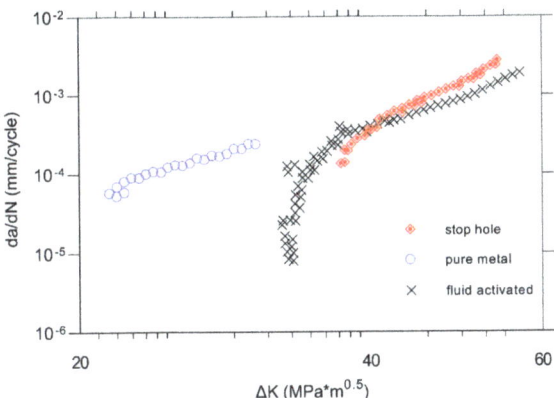

Figure 8. Comparison of fatigue crack growth rates for pure metallic specimens (stop hole, CT specimen with injected fluid).

Noticeable is the high deceleration effect caused by injected fluid—much stronger than in stop hole technique. The crack growth rate decreases at least ten times in the initial injection. As the crack length increases (and thus injections were interrupted, the crack was dried, for approx. $\Delta K = 37-38$ MPa*m$^{0.5}$), the crack growth rate (characterized by slope of the linear part of KFFD) was similar to the initial slope and initial FCGR—like that obtained for pure metal (large dots in Figure 8). Based on the above, it was decided to analyze fatigue fracture surfaces using light and scanning electron (SEM) microscopy in order to analyze the crack growth mechanism in both cases (with and without fluid) due to the observed significant retardation effect of fatigue crack growth rate -before and after of the injection. Additionally, the same type of observations was done for the "stop hole" specimen. In Figures 9 and 10 are presented fatigue crack surfaces in the vicinity of drilled stop-hole in a macroscopic view. Detailed SEM analysis before drilling hole (crack tip position 3 mm after pre-crack) does not differ from the original crack path in this steel [24]. As the crack grows (under constant load amplitude) an increasing number of secondary cracks is observed as a natural result of increased K_I values under tensile crack growth mode—Figure 11. After drilling the hole (Figure 12) 75,000 cycles to 110,000 cycles were required for re-initiation of fatigue crack growth. After that, the crack starts to grow similarly as in pure CT specimen under the tensile mode, which is reflected in Figure 13 on fractograms with a large number of secondary cracks and noticeable fatigue striations.

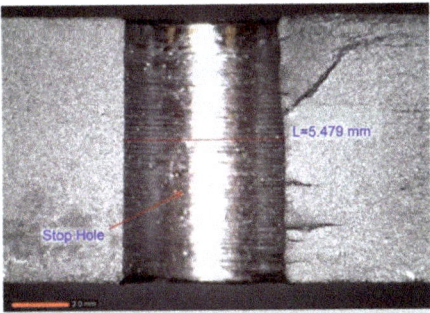

Figure 9. Specimen after stop hole technique testing procedure with re-initiated crack after the hole (crack growth direction from left to right).

Figure 10. Initial fatigue crack path (15 mm) of the tested specimen with stop hole ($\Delta K = 24$ MPa $\times$ m$^{0.5}$), crack growth direction from bottom to top.

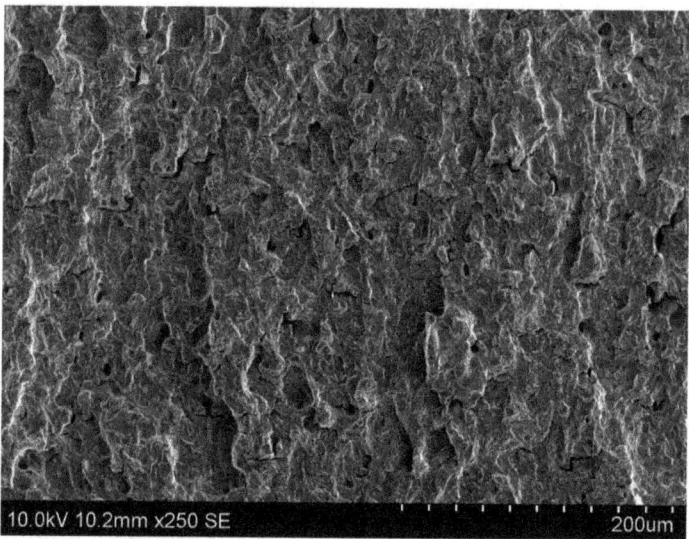

Figure 11. Initial fatigue crack path (22 mm) of the tested specimen with stop hole ($\Delta K = 34.8$ MPa $\times$ m$^{0.5}$), crack growth direction from bottom to top, fracture surface located close to drilled hole.

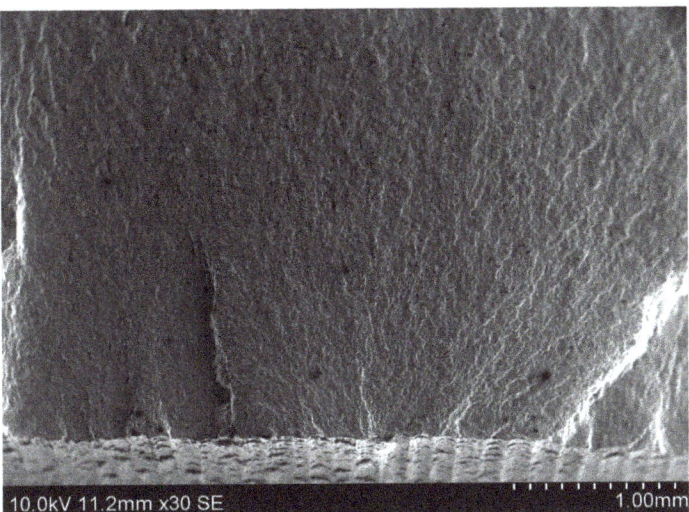

Figure 12. Initial fatigue crack path (24.3 mm) of the tested specimen with stop hole (ΔK = 40 MPa × $m^{0.5}$), crack growth direction from bottom to top, fracture surface located close to drilled hole with reinitiated fatigue crack.

Figure 13. Initial fatigue crack path (24.5 mm) of the tested specimen with stop hole (ΔK = 40.4 MPa × $m^{0.5}$), crack growth direction from bottom to top, fracture surface located close to the drilled hole with reinitiated fatigue crack—noticeable fatigue striations.

On the contrary to the typical crack growth mechanism in specimen tested in fluid injections—Figure 14—an etched fracture surface was observed due to a possible mechanism of interaction fluid with crack surfaces and chemical reactions [19]. As a result of interaction between the fluid and crack surfaces, a solid product of substantial volume appears, which fills the crack cavity. The chemical mechanism on which this method is based is similar to intrinsic crack closure caused by products of interaction between metal and humid air or the corrosive environment due to fretting corrosion. The active fluid components interact with ferrous ions and form insoluble complexes. When steel is exposed to an electrolyte (in this case to fluid), metal ions leave the lattice and enter the electrolyte as ferrous ions. Ferrous ions and active fluid ions react to form insoluble compounds, chelate Fe(II).

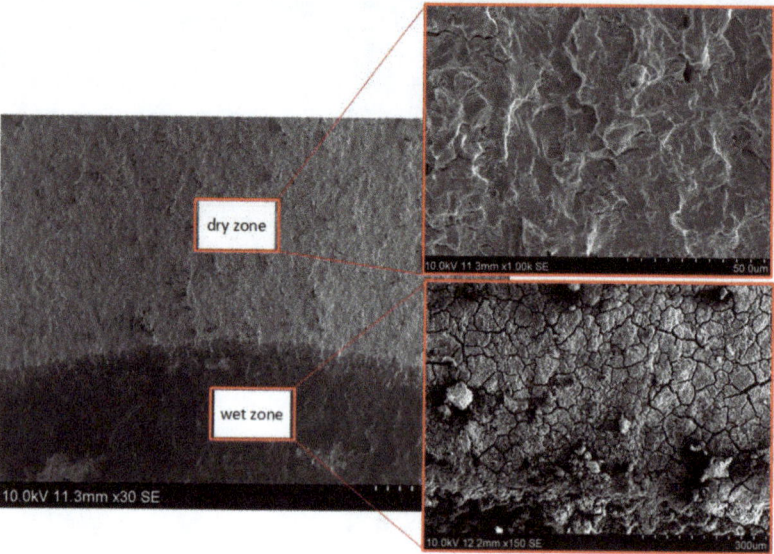

Figure 14. SEM macroscopic view on the "wet" and "dry" zone of crack propagation mechanism in specimens tested with fluid injections.

However, natural crack closure is peculiar to fatigue crack growth at low ΔK. Therefore, the task consisted of searching for such substance that would rapidly provide much more intensive interaction with the metallic specimen's crack surfaces.

The fatigue fracture surface in Figure 10 is mainly shaped by noticeable longitudinal ridges (normal to the direction of maximum tensile stress) and facets associated with various pearlite colony orientations.

With increasing crack length, fatigue fracture surface (Figure 11) is mainly shaped by numerous secondary cracks resulting from increasing tensile stress in front of the growing crack for higher values of stress intensity factor.

Large ridges mostly shaped macroscopic view on close area to drilled hole—called "re-initiation region" (Figure 12). The fracture surface for increased crack length (Figure 13) was characterized by microcracks in the interphase zone (plates of cementite and ferrite) on the background of fatigue striations.

After testing in environmental—fluid—conditions, the region on the border between dry and wet crack was particularly interesting for SEM investigations. In Figure 14, a macroscopic view of the crack front between "wet" and "dry" zone (approx. after 150,000 cycles) is shown with marked by frames microscopic SEM images.

As it was expected in the region without fluid influence "dry zone"—Figure 14—crack growth mechanism was typical for this steel as it was evidenced in previous specimens (i.e., stop hole technique) and results in a similar final slope of the da/dN–ΔK diagram. This region is also shaped mainly by the number of secondary cracks and fatigue striations. The "wet zone" is characterized by many metal–fluid reaction products and wear parts of the fracture surface as a consequence of the oxidization and finally artificial crack closure effect. Initially, based on PICC (Plasticity Induced Crack Closure). Elber [22] defined the effective stress intensity factor range as:

$$\Delta K_{eff} = K_{max} - K_{op}, \quad (2)$$

where K_{max} is the maximal value of stress intensity factor; K_{op} is the stress intensity factor at crack opening during fatigue cycle. Finally, the closure parameter U can be defined as:

$$U = \frac{\Delta K_{eff}}{\Delta K_{app}}. \quad (3)$$

For the tested specimen, crack closure parameter U was calculated based on LQSM (Linear Quadratic Spline Method) successfully applied in previous Authors' papers [26]. As observed from fracture surfaces, many fluid chemical reactions products caused a significant drop of U-value, which constitutes effective crack driving force and finally deceleration of fatigue crack growth. A significant decrease of U-parameter was observed after fluid injection—Figure 15. This fact may explain the combined physico-chemical artificial crack closure effect (PCMACC).

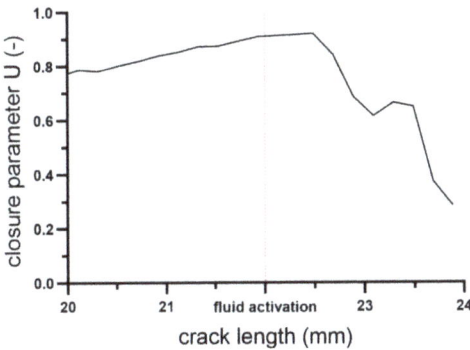

Figure 15. Elber closure parameter variation for "fluid" specimen before and after injection (including 2 mm crack length increment).

4. Conclusions

Comparison of fatigue lifetime curves for all specimens tested in the same loading conditions shown that a noticeable comparable effect was obtained for the designed stop hole technique and composite CFRP patches.

Moreover, a comparison of FCGR methods shown a noticeable higher deceleration effect caused by injected fluid than in the stop hole technique. The crack growth rate decreases at least ten times in the initial injection.

The detailed SEM analysis of the stop hole technique's fractogram showed no differences between paths of this sample and the raw steel before drilling the hole. After drilling the hole, crack starts to grow in a similar manner as in pure CT specimen under the tensile mode, which is reflected on fractograms with a large number of secondary cracks and noticeable fatigue striations.

On the contrary to the typical crack growth mechanism in (raw specimen), in the specimen tested with fluid injections, an etched fracture surface was observed. This chemical treatment triggers the mechanism on which this method is based and which is similar to intrinsic crack closure caused by products of interaction between metal and humid air or the corrosive environment due to fretting corrosion.

Based on experimental results presented in this paper, as well as in previous paper devoted to low-carbon steel [19], it can be concluded that the fluid should be thin enough to fall into the crack easily and fill it. What is more, the chemical substance should be dissoluble to concentrations that effectively delay crack growth. It is worth noting that this effect should be tested on a wider group of materials to validate its usefulness in engineering practice combined with analytical modelling of crack growth retardation effect.

Author Contributions: Conceptualization, H.N. and J.A.F.O.C.; methodology, G.L.; software, S.D.; validation, G.L., J.M.D.A.F. and R.M.; formal analysis, G.L.; investigation, G.L., B.B., J.M.D.A.F. and O.Z.; data curation, R.M.; writing—original draft preparation, G.L., J.M.D.A.F. and R.M.; writing—review and editing, G.L.; visualization, S.D.; supervision, J.A.F.O.C.; project administration, O.Z.; funding acquisition, G.L. All authors have read and agreed to the published version of the manuscript.

Funding: The project was supported in part by the Polish National Agency for Academic Exchange (Polish–Ukrainian bilateral agreement) grant number PPN/BUA/2019/1/00086.

Institutional Review Board Statement: Not applicable.

Informed Consent Statement: Not applicable.

Data Availability Statement: Not applicable.

Conflicts of Interest: The authors declare no conflict of interest.

References

1. Zerbst, U.; Mädler, K.; Hintze, H. Fracture mechanics in railway applications—An overview. *Eng. Fract. Mech.* **2005**, *72*, 163–194. [CrossRef]
2. Clayton, P. Predicting the wear of rails on curves from laboratory data. *Wear* **1995**, *181*, 11–19. [CrossRef]
3. Clayton, P.; Jin, N. Unlubricated sliding and rolling/sliding wear behavior of continuously cooled, low/medium carbon bainitic steels. *Wear* **1996**, *200*, 74–82. [CrossRef]
4. Muster, H.; Schmedders, H.; Wick, K.; Pradier, H. Rail rolling contact fatigue. The performance of naturally hard and head-hardened rails in track. *Wear* **1996**, *191*, 54–64. [CrossRef]
5. Heyder, R.; Girsch, G. Testing of HSH rails in high-speed tracks to minimize rail damage. *Wear* **2005**, *258*, 1014–1021. [CrossRef]
6. Wetscher, F.; Stock, R.; Pippan, R. Changes in the mechanical properties of a pearlitic steel due to large shear deformation. *Mater. Sci. Eng.* **2007**, *445*, 237–243. [CrossRef]
7. Hassani, A.; Ravaee, R. Characterization of transverse crack and crack growth in a railway rail. *Iran. J. Mater. Sci. Eng.* **2008**, *5*, 22–31.
8. Hohenwarter, A.; Taylor, A.; Stock, R.; Pippan, R. Effect of large shear deformations on the fracture behavior of a fully pearlitic steel. *Met. Mater. Trans. A* **2011**, *42*, 1609–1618. [CrossRef]
9. Lesiuk, G.; Katkowski, M.; Correia, J.; de Jesus, A.M.; Blazejewski, W. Fatigue crack growth rate in CFRP reinforced constructional old steel. *Int. J. Struct. Integr.* **2018**, *9*, 381–395. [CrossRef]
10. Yu, Q.Q.; Wu, Y.F. Fatigue retrofitting of cracked steel beams with CFRP laminates. *Compos. Struct.* **2018**, *192*, 232–244. [CrossRef]
11. Emdad, R.; Al-Mahaidi, R. Effect of prestressed CFRP patches on crack growth of centre-notched steel plates. *Compos. Struct.* **2015**, *123*, 109–122. [CrossRef]
12. Hosseini, A.; Ghafoori, E.; Motavalli, M.; Nussbaumer, A.; Zhao, X.L. Mode I fatigue crack arrest in tensile steel members using prestressed CFRP plates. *Compos. Struct.* **2017**, *178*, 119–134. [CrossRef]
13. Yu, Q.Q.; Gao, R.X.; Gu, X.L.; Zhao, X.L.; Chen, T. Bond behavior of CFRP-steel double-lap joints exposed to marine atmosphere and fatigue loading. *Eng. Struct.* **2018**, *175*, 76–85. [CrossRef]
14. Jin, K.; Chen, K.; Luo, X.; Tao, J. Fatigue crack growth and delamination mechanisms of Ti/CFRP fibre metal laminates at high temperatures. *Fatigue Fract. Eng. Mater. Struct.* **2020**, *43*, 1115–1125. [CrossRef]
15. Borrie, D.; Raman, X.L.Z.R.S.; Bai, Y. CFRP strengthened pre-cracked steel plates protected with chemical silane exposed to extreme marine environments. In *International Symposium on Fiber Reinforced Polymers for Reinforced Concrete Structures (FRPRCS) the Asia-Pacific conference on fiber reinforced polymers in structures (APFIS)*; Southeast University: Dhaka, Bangladesh, 2015; pp. 1–6.
16. Razavi, S.M.J.; Ayatollahi, M.R.; Sommitsch, C.; Moser, C. Retardation of fatigue crack growth in high strength steel S690 using a modified stop-hole technique. *Eng. Fract. Mech.* **2017**, *169*, 226–237. [CrossRef]
17. Ayatollahi, M.R.; Razavi, S.M.J.; Chamani, H.R. Fatigue life extension by crack repair using stop-hole technique under pure mode-I and pure mode-II loading conditions. *Procedia Eng.* **2014**, *74*, 18–21. [CrossRef]
18. Wu, H.; Imad, A.; Benseddiq, N.; de Castro, J.T.P.; Meggiolaro, M.A. On the prediction of the residual fatigue life of cracked structures repaired by the stop-hole method. *Int. J. Fatigue* **2010**, *32*, 670–677. [CrossRef]
19. Khaburskyi, Y.; Slobodyan, Z.; Hredil, M.; Nykyforchyn, H. Effective method for fatigue crack arrest in structural steels based on artificial creation of crack closure effect. *Int. J. Fatigue* **2019**, *127*, 217–221. [CrossRef]
20. Borrego, L.P.; Ferreira, J.M.; Da Cruz, J.P.; Costa, J.M. Evaluation of overload effects on fatigue crack growth and closure. *Eng. Fract. Mech.* **2003**, *70*, 1379–1397. [CrossRef]

21. Maierhofer, J.; Simunek, D.; Gänser, H.P.; Pippan, R. Oxide induced crack closure in the near threshold regime: The effect of oxide debris release. *Int. J. Fatigue* **2018**, *117*, 21–26. [CrossRef]
22. Elber, W. Fatigue crack closure under cyclic tension. *Eng. Fract. Mech.* **1970**, *2*, 37–45.
23. Nykyforchyn, H.; Pustovyi, V.; Slobodyan, Z.; Khaburskyi, Y.; Barna, R.; Zvirko, O.; Kret, N. The Method of Fatigue Crack Growth Arrest. Patent of Ukraine No.128514, 15 September 2018. (In Ukrainian).
24. Lesiuk, G.; Smolnicki, M.; Mech, R.; Zięty, A.; Fragassa, C. Analysis of fatigue crack growth under mixed mode (I+ II) loading conditions in rail steel using CTS specimen. *Eng. Fail. Anal.* **2020**, *109*, 104354. [CrossRef]
25. Lesiuk, G.; Pedrosa, B.A.; Zięty, A.; Błażejewski, W.; Correia, J.A.; De Jesus, A.M.; Fragassa, C. Minimal invasive diagnostic capabilities and effectiveness of CFRP-Patches repairs in long-term operated metals. *Metals* **2020**, *10*, 984. [CrossRef]
26. Lesiuk, G.; Szata, M.; Correia, J.A.; De Jesus, A.M.P.; Berto, F. Kinetics of fatigue crack growth and crack closure effect in long term operating steel manufactured at the turn of the 19th and 20th centuries. *Eng. Fract. Mech.* **2017**, *185*, 160–174. [CrossRef]

Article

Fatigue Crack Growth Behaviour and Role of Roughness-Induced Crack Closure in CP Ti: Stress Amplitude Dependence

Mansur Ahmed [1,*], Md. Saiful Islam [2], Shuo Yin [1], Richard Coull [1] and Dariusz Rozumek [3]

[1] Department of Mechanical, Manufacturing and Biomedical Engineering, Trinity College Dublin, The University of Dublin, D02 DA31 Dublin, Ireland; yins@tcd.ie (S.Y.); rncoull@live.com (R.C.)
[2] Department of Glass and Ceramic Engineering, Bangladesh University of Engineering and Technology, Dhaka 1000, Bangladesh; mdsaiful@gce.buet.ac.bd
[3] Faculty of Mechanical Engineering, Opole University of Technology, Mikolajczyka 5, 45-271 Opole, Poland; d.rozumek@po.edu.pl
* Correspondence: maahmed@tcd.ie or ma960@uowmail.edu.au; Tel.: +353-1-896-2396

Citation: Ahmed, M.; Islam, M.S.; Yin, S.; Coull, R.; Rozumek, D. Fatigue Crack Growth Behaviour and Role of Roughness-Induced Crack Closure in CP Ti: Stress Amplitude Dependence. *Metals* **2021**, *11*, 1656. https://doi.org/10.3390/met11101656

Academic Editor: Denis Benasciutti

Received: 16 September 2021
Accepted: 18 October 2021
Published: 19 October 2021

Publisher's Note: MDPI stays neutral with regard to jurisdictional claims in published maps and institutional affiliations.

Copyright: © 2021 by the authors. Licensee MDPI, Basel, Switzerland. This article is an open access article distributed under the terms and conditions of the Creative Commons Attribution (CC BY) license (https://creativecommons.org/licenses/by/4.0/).

Abstract: This paper investigated the fatigue crack propagation mechanism of CP Ti at various stress amplitudes (175, 200, 227 MPa). One single crack at 175 MPa and three main cracks via sub-crack coalescence at 227 MPa were found to be responsible for fatigue failure. Crack deflection and crack branching that cause roughness-induced crack closure (RICC) appeared at all studied stress amplitudes; hence, RICC at various stages of crack propagation (100, 300 and 500 µm) could be quantitatively calculated. Noticeably, a lower RICC at higher stress amplitudes (227 MPa) for fatigue cracks longer than 100 µm was found than for those at 175 MPa. This caused the variation in crack growth rates in the studied conditions.

Keywords: CP Ti; stress amplitude; fatigue crack propagation; crack growth rate; roughness-induced crack closure

1. Introduction

Commercially pure titanium (CP Ti) possesses high ductility as well as excellent corrosion resistance and biocompatibility properties; hence, it has been used in the chemical and biomedical industries, especially in reactor container in chemical plants, and in power station heat exchangers [1]. In these environments, cyclic loading is applied to the components. Therefore, investigation of the fatigue behaviour of CP Ti becomes an important research subject. Fatigue crack initiation, growth, closure and fractography are critical features describing fatigue behaviour. A fatigue crack path may be a powerful resource determining all the aforementioned features. For instance, roughness-induced crack closure (RICC) is attributed to crack path deflection, especially near the threshold range, at which a serrated or zigzag crack path is induced by microstructure-sensitive crack growth [2–4]. The tilt angle of the crack path in 2124 Al alloy is reported to be a key controlling factor for RICC [5]. Wang and Müller [2,6] reported that RICC occurs due to a serrated crack path, which significantly affects crack growth rates in Ti-2.5 Cu (wt%) and TIMETAL 1100 alloys. Fatigue crack growth rates are reported to be decreased by crack path deflection [3,4,7], as the crack path causes a direct reduction of the local driving force for crack propagation and an increase in the total crack path length, which results in lower crack growth rates and induces RICC. Antunes et al. [8] mentioned that cracks with larger path deflection may result in higher mode II displacement between the two fracture surfaces, causing higher crack closure stress intensity [9]. Ding et al. [10] correlated the fatigue crack profile with fatigue crack growth rate in Ti–6Al–4V and Ti–4.5Al–3V–2Mo–2Fe alloys. Okayasu et al. [11] investigated the influence of fatigue loading conditions such as, stress amplitude and stress ratio on the contact features of fracture surfaces in annealed Carbon Steel via two methods: (1) Collecting fracture debris fallen from the crack surfaces; and (2) Observing the fracture surface directly through the replica technique. They showed that the fatigue stress amplitude and stress ratio played major factors in determining the contact status between the mating fracture surfaces, e.g., a larger stress amplitude and smaller stress ratio lead to stronger fracture surface contact or interaction. Student et al. [12] studied the role of fracture surface roughness on crack closure in long-term exploited heat-resistant steel. They reported that shear processes at the tip of a fatigue crack significantly affect crack closure and contribute to the roughness of the fracture surface.

The above discussions evidently show that fatigue crack path can be effectively used to investigate numerous phenomena. Unfortunately, study on the fatigue behaviour of CP Ti related to crack path and its role in detecting fatigue phenomena such as RICC is totally lacking in the literature. To fill this gap, the fatigue crack growth mechanism and corresponding RICC from fatigue crack path in CP Ti has been revealed here. This will further help in understanding fatigue crack growth at the small crack regime and the role of RICC in such cracking.

2. Materials and Methods

The composition of the studied CP Ti is given in Table 1. A vacuum furnace heat-treatment of the as received hot rolled sample was performed at 700 °C for 30 min for obtaining equiaxed α grains. Following heat treatment, samples were machined in order to prepare fatigue specimen with the dimensions shown in Figure 1. Because the surface condition of the specimen in fatigue testing is very sensitive, the samples were then carefully mechanically polished using various grades of SiC emery papers followed by colloidal silica with hydrogen peroxide solution. The polished surfaces were then etched using Kroll's reagent ($3HF:6HNO_3:91H_2O$) to reveal the microstructure.

Table 1. Chemical composition (in wt%) of the CP Ti.

Ti	O	H	Fe
Balance	0.078	0.005	0.026

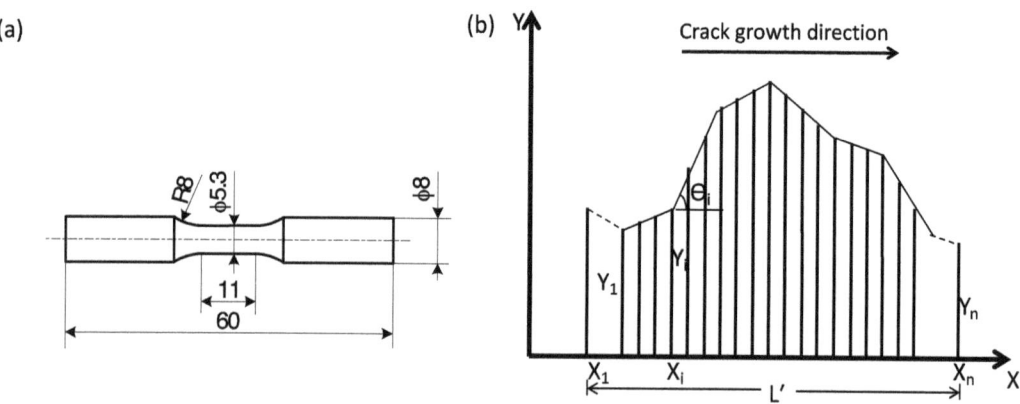

Figure 1. (a) Schematic of sample with the dimensions used for the rotating bending test (dimensions in mm); (b) Y-coordinates along the crack path profile with equidistant spacing.

Tensile tests of CP Ti samples were performed employing an Instron Universal Tensile Machine. The dimensions of the samples were in accordance with the ASTM E8. For each condition, three samples are tested, at room temperature and at a strain rate of 1 mm/min.

An Ono-type rotating bending fatigue machine was used for fatigue tests, at a frequency of 30 Hz. A stress interval of 25 MPa was chosen, from the fatigue limit to a stress at which the distinction of the stress amplitude effect could be numerated. Consequently, fatigue tests at the stresses of 175, 200 and 227 MPa were conducted. The stress ratio (R) was −1 with sinusoidal wave form. All tests were conducted at temperature from 15–20 °C. During testing, a cooling fan was used to cool the specimens, as heat production affects the deformation mechanism of Ti. During fatigue testing, the replica technique was intermittently used at different cycles to trace out crack initiation and propagation; this technique involved immersing replica films into methyl acetate solution and subsequently pasting them onto the specimen surface. To then acquire the image from the replica sheet, an optical microscope (VHX-2000 series, Keyence, Osaka, Japan) was used. One sample of each condition was investigated. Following the fatigue tests, fractographic analysis was conducted using a JEOL IT-300 (JEOL, Tokyo, Janpa) scanning electron microscope (SEM) at an acceleration voltage of 30 kV.

Roughness parameters were quantitatively measured using the surface crack paths along the fractured specimens. Two roughness parameters, (i) linear roughness parameter (R_L) and (ii) arithmetic mean deflection angle ($\bar{\theta}$), were evaluated using the equidistant spacing method [3,4,13]. Figure 1b displays the equidistant spacing method used to calculate these roughness parameters. The details of this method can be found elsewhere [3].

The linear roughness parameter, R_L of a crack profile is defined as follows:

$$R_L = L/L' \qquad (1)$$

where L and L' correspond to the true length and projected length of the crack profile, respectively. The arithmetic mean deflection angle of a crack profile is given below:

$$\bar{\theta} = \frac{1}{n}\sum_{i=1}^{n}|\theta_i| \qquad (2)$$

where θ_i represents the angle between the profile element and X-coordinate axis and can have either a positive or a negative value ($90° \leq \theta \leq -90°$), depending on the crack profile.

3. Results

3.1. Initial Microstructure, and Tensile and Fatigue Properties

Figure 2a shows the initial microstructure of the studied material. Uniformly distributed (hexagonal closed packed) α grains with an average diameter of ~35 μm can be observed. The size of the α grains was calculated using Image J. The engineering stress–strain curve plotted from the tensile test is shown in Figure 2b. Beyond the yielding point, work-hardening is followed by work-softening along the stress–strain curve. The yield strength (YS), ultimate tensile strength (UTS) and total elongation (El.) of the studied material were measured to be 293 ± 12 MPa, 383 ± 7 MPa and 67 ± 1%, respectively. Table 2 shows the fatigue test results at various stress amplitudes. As expected, the total life cycle was found to be reduced with increasing stress amplitude. For instance, the sample tested at 175 MPa survived for 6.94×10^5 cycles, while the sample tested at 227 MPa only lasted for 1.05×10^5 cycles. At the intermediate stress amplitude of 200 MPa, the sample failed after reaching 3.4×10^5 cycles.

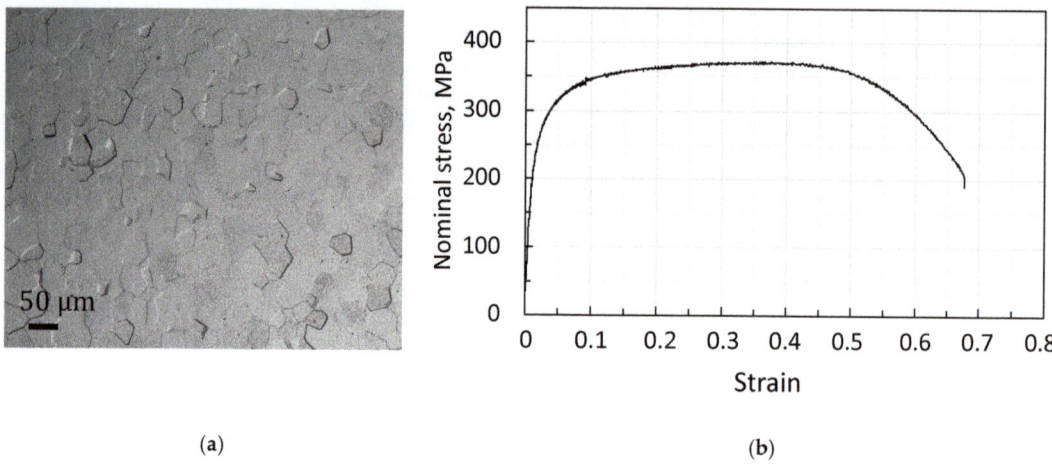

Figure 2. (a) Initial microstructure; (b) Representative stress–strain curve of the studied CP Ti.

Table 2. Summary of fatigue tests results obtained in this study.

Stress Amplitude, MPa	Crack Initiation Cycle ($\times 10^5$)	Total Life Cycle ($\times 10^5$)
175	1.0–2.0	6.94
200	0.5–1.0	3.40
227	0.2–0.3	1.05

3.2. Fatigue Crack Nucleation and Propagation, Propagation Rate and Fractography

Figure 3 shows optical micrographs of the fatigue surface crack and its surroundings at various cycles at 175 MPa. The microstructure in Figure 3a shows an image taken before the test corresponding to a region where the main fatigue crack initiated. After 2×10^5 cycles, a micro-sized crack located inside an α grain can be seen (Figure 3b). This indicates that the crack initiated between 1×10^5–2×10^5 cycles, as no crack was found after 1×10^5 cycle (not shown here). Therefore, it is evident that most of the fatigue lifecycle was consumed by fatigue crack propagation considering total fatigue life (6.94×10^5 cycles). Crack propagation and its features after 4×10^5 cycles can be observed in Figure 3c, and was predominantly transgranular in nature and thus deflected by almost every grain. Following 6×10^5 cycles (Figure 3d), crack propagation was incremental, continuing in transgranular mode. Some important fatigue crack propagation features, such as crack branching (Figure 3f,h) and fine scale zig-zag (Figure 3e–h) can also be seen. Some lines, presumably slip bands and/or deformation twinning, according to [14], appear in some grains (Figure 3f,h). Interactions between the crack and those lines confirm that cracking propagated along or across these slip bands/deformation twinning. Ismarrubie and Sugano [15] have also reported such lines belonging to slip bands. While crack branching in Figure 3f is transgranular, intergranular cracking also appears in Figure 3h.

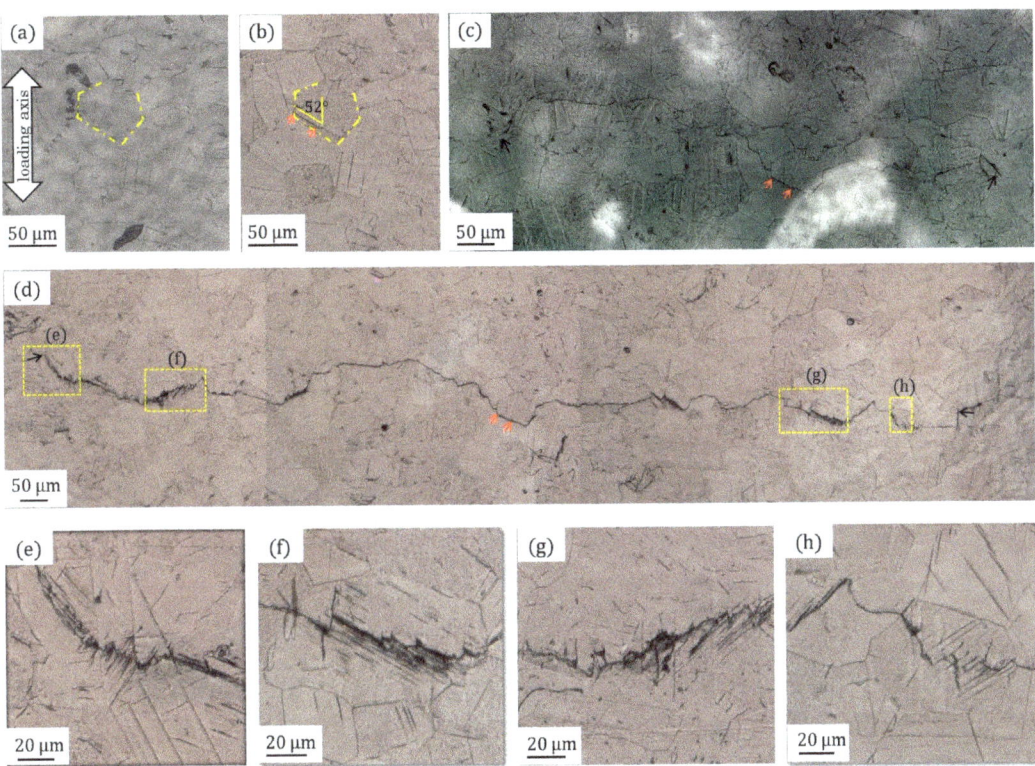

Figure 3. At 175 MPa: Images showing surface crack paths at different cycles: (**a**) Initial condition, i.e., N = 0 cycle; (**b**) After N = 2 × 10^5 cycles; (**c**) After N = 4 × 10^5 cycles; (**d**) After N = 6 × 10^5 cycles. The free-shape object in (**a**) and (**b**) indicates a particular area where the crack started. The red arrows illustrate crack initiation sites, and black arrows delineate the crack tips. Examples of fine zig-zag (**e**,**g**), crack branching (**f**,**g**), and crack interactions with slip bands (**e**,**h**).

Figure 4 delineates the initiation and subsequent propagation of the main crack at various cycles at 227 MPa. Figure 4a illustrates the main fatigue crack, with a zig-zag pattern, after 1 × 10^5 cycles. This crack has been sectioned to understand its propagation mechanism. There are three sub-cracks labeled 1–3 connected to the main crack. Sub-cracks were labeled according to their connecting sequence with the main crack. Therefore, it can be seen that the fatigue crack of CP Ti tested at higher stress amplitude (227 MPa) grew by coalescing sub-cracks. Figure 4b shows the main crack after 0.4 × 10^5 cycles, with the crack initiation site marked by red arrows. It is worth mentioning that the crack initiated between 0.2 × 10^5–0.3 × 10^5 cycles, as there was no crack at 0.2 × 10^5 cycles (not shown here). Similar to the sample tested at 175 MPa, this condition also consumed majority of the total cycle of crack growth. Figure 4c–e show micrographs after 0.5 × 10^5, 0.7 × 10^5 and 0.9 × 10^5 cycles, respectively, where pre-coalescence of the sub-cracks with the main crack are shown. Unlike the crack at 175 MPa, the crack at 227 MPa preferably propagated via coalescing sub-cracks. This is the first time we saw such a difference in crack propagation mechanism with respect to the stress amplitude in CP Ti. Of the fatigue characteristics, a zig-zag nature and fine-scale crack branching are also visible under this condition. As at 175 MPa, a transgranular fracture mode was also predominant in this case. It is worth mentioning that crack propagation behavior at 200 MPa was identical to that at 227 MPa.

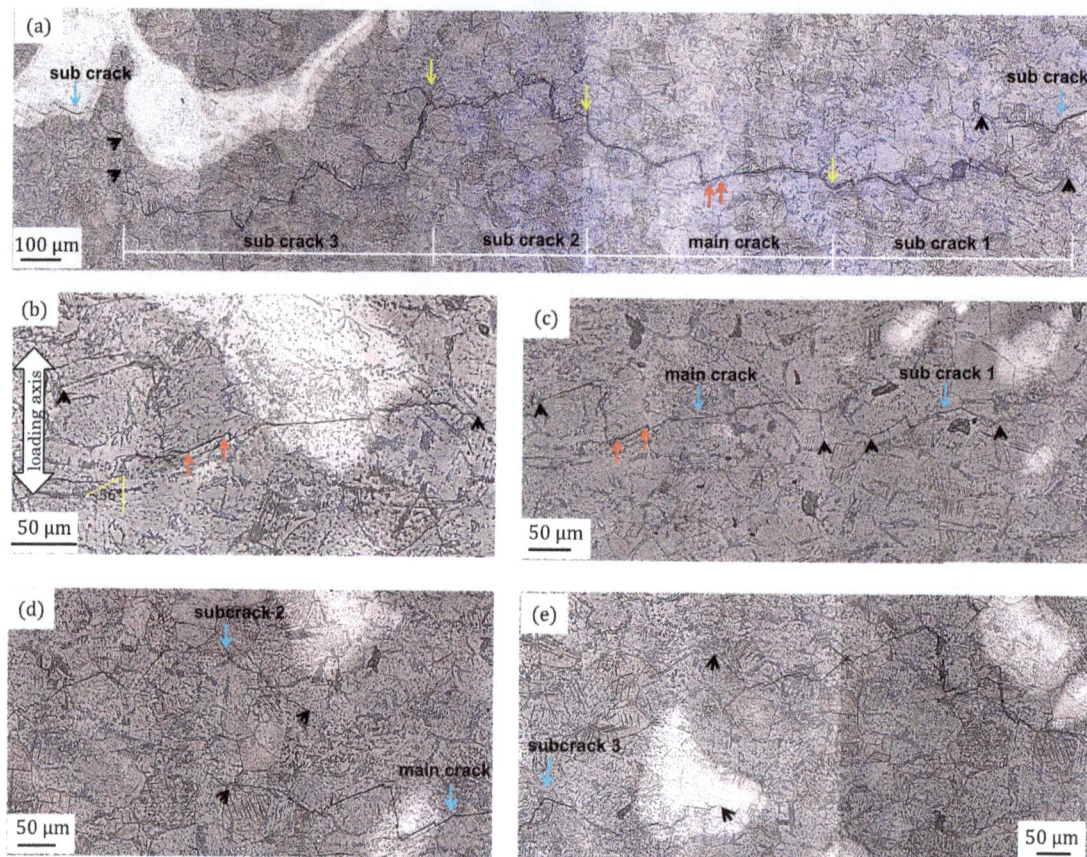

Figure 4. At 227 MPa: Images displaying crack paths at different cycles. After (**a**) $N = 1 \times 10^5$ cycles; (**b**) $N = 0.4 \times 10^5$ cycles; (**c**) $N = 0.5 \times 10^5$ cycles; (**d**) $N = 0.7 \times 10^5$ cycles; (**e**) $N = 0.9 \times 10^5$ cycles. Red arrows show the crack initiation line, while black arrows indicate crack tips; (**c**) shows the main crack and sub-crack 1 before coalescing, while the main crack and sub-crack 2 can be seen in (**d**); (**e**) shows the main crack and sub-crack 3 prior to coalescence. The yellow arrows in (**a**) indicate the junctions of crack coalescence.

Figure 5 shows the relationship between crack length and number of cycles, as well as crack propagation rate with respect to the crack length. These data have been calculated based on the crack paths shown in Figures 3 and 4. While the crack length in the 175 MPa sample grows steadily, crack length at 227 MPa increases abruptly after around 300 μm (Figure 5a). On the other hand, the crack growth rate in the 227 MPa sample is significantly higher than that of the 175 MPa sample for cracks longer than 100 μm. Figure 6 shows the fracture surfaces of the samples tested at various stress amplitudes. Three main features, including (i) crack initiation site (marked with green boxes), (ii) crack propagation, and (iii) the final fracture as dimples can be observed in each sample. Higher magnification images of the crack initiation sites are shown in Figure 6d–f. Interestingly, the position of the dimples progressively moves toward centre of the sample with increased stress amplitude. This corresponds to the number of cracks responsible for fracture with stress amplitude; for instance, one single crack, two cracks and three cracks were responsible for fracture of samples tested at 175, 200 and 227 MPa, respectively.

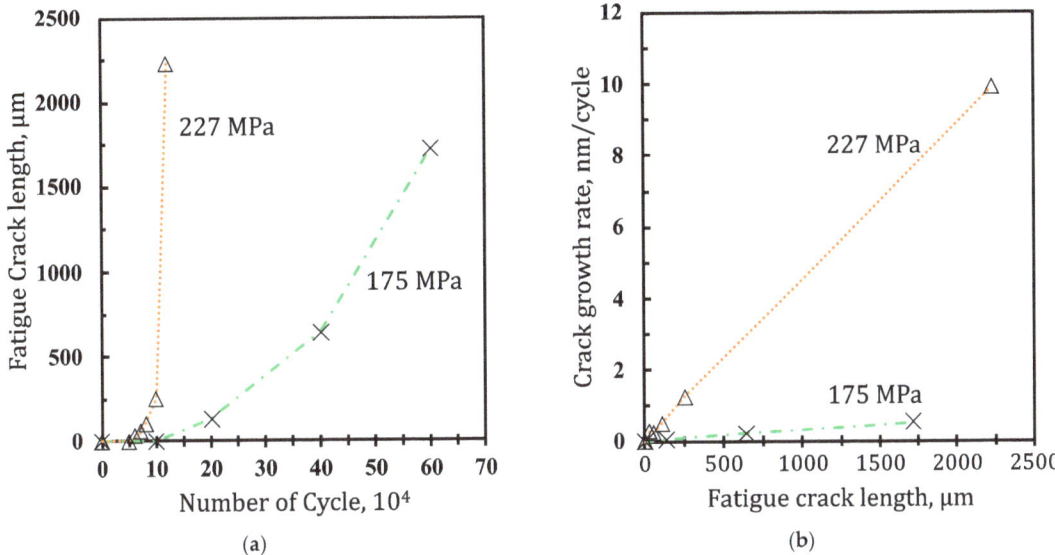

Figure 5. (**a**) Relationship between fatigue crack length and number of cycles; (**b**) Crack growth rate with respect to fatigue crack length.

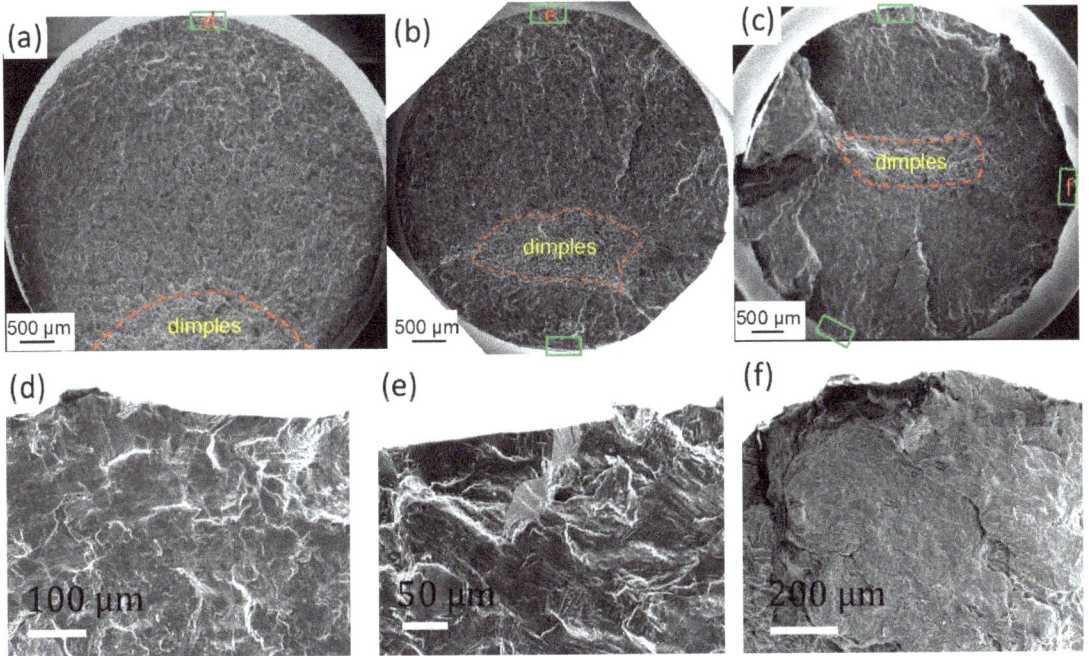

Figure 6. Fractography of the samples tested at (**a**) 175, (**b**) 200, and (**c**) 227 MPa. Green boxes indicate the nucleation sites of the main cracks. Crack nucleation sites for (**d**) 175, (**e**) 200, and (**f**) 227 MPa stress conditions.

4. Discussion

4.1. Fatigue Crack Initiation and Propagation

Fatigue cracks at various stress amplitudes evidently initiated from the surfaces of the specimens (Figure 6d–f). This is consistent with earlier claims that fatigue cracks in Ti and its alloys tend to initiate from surface in the case of continuous cyclic loading if the surface is free of residual stress [16]. In this study, mechanical polishing followed by chemical etching was performed in order to ensure a residual stress-free surface so that crack would start from specimen surface. The measured angle of ~50° (Figures 3b and 4b) between crack initiation and the loading axis is close to that of the maximum critical resolved shear stress (45°). The cracks thereafter propagated along a direction of approximately 70° with respect to the loading axis, which is corresponded to mixed mode I and mode II crack growth. Close examination of crack path shows that a significant portion of the cracks propagated as zig-zag where the crack paths moved along or across slip bands/deformation twinning in short distances (Figure 4e–h). Such zig-zag phenomenon has been attributed to the alternative branching mode I and mode II in forged VT3-1 alloy [17]. Some portion of the crack propagated along the direction perpendicular to the loading axis, corresponding to mode II. It has been mentioned in [18,19] that a shift in direction perpendicular to the specimen may sometimes appear in crack branching. This is consistent with the results shown in Figure 4f. The nature of the crack path is likely not dependent on the stress amplitude, while the number of cracks causing failure is highly dependent upon the stress amplitude. At 175 MPa, a single crack was responsible for failure, whereas at higher stress amplitudes the number of cracks that are responsible for failure increases. For instance, two cracks and three cracks nucleated at different stages in samples tested at 200 and 227 MPa (Figure 3b,c), respectively. Each of the main crack propagated via the sub-crack coalescence mechanism (Figure 5). Therefore, it can be claimed that fatigue crack nucleation and their propagation in CP Ti are largely dependent on the stress amplitude.

4.2. Role of Roughness Induced Crack Closure (RICC)

It is established that crack tortuosity, crack branching or their combination induce crack closure, as they promote higher roughness [19]. In this study, we have seen crack deflections with a zig-zag nature, crack branching, and a directional shift perpendicular to the specimen surrounding crack branching. Therefore, it is assumed that RICC has played a role in crack propagation. As such, the RICC for 175 and 227 MPa samples has been calculated; the stress amplitude dependence of RICC is discussed below.

A model proposed by Pokluda and Pippan is used to quantitatively measure RICC [20]. The total maximum level of RICC, $\left(\frac{\delta_{cl}}{\delta_{max}}\right)_{RICC}$, can be expressed as follows:

$$\left(\frac{\delta_{cl}}{\delta_{max}}\right)_{RICC} = C\eta\sqrt{(R_\theta^2 - 1)} + \frac{3\eta(R_\theta - 1)}{[\sqrt{6} + 3(R_\theta - 1)]} \tag{3}$$

where $C \approx 10^{-1}$ is a dimensionless constant independent of material, $R_\theta = \cos^{-1}(\overline{\theta})$, is the arithmetic mean of the angle that dictates the crack deflection with crack propagation (Figure 7b). η is a parameter that strongly depends on the size ratio, SR = dm/rp where dm is the mean grain size and rp is the static plastic zone size. Static plastic zone size varies depending on the maximal applied intensity factor as a function of the applied stress amplitude and the crack length. Therefore, the value of η differs for different stress amplitudes and, moreover, significantly changes during crack propagation. To assess the values of η at various crack lengths, the following equation [21] is applied:

$$\eta = \exp\left[-(0.886 \, rp/dm)^{2.2}\right] \tag{4}$$

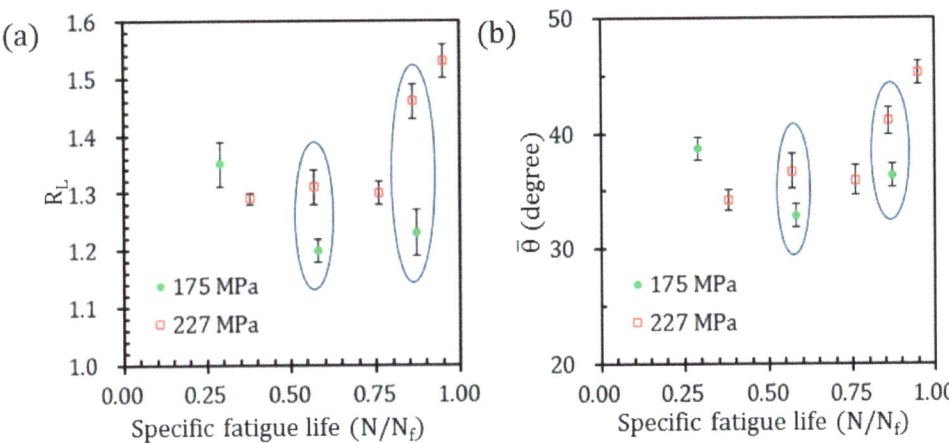

Figure 7. Roughness parameters of samples tested at the stress of 175 and 227 MPa based on the surface crack path profile: (**a**) Linear roughness parameter, R_L; (**b**) Arithmetic mean of the deflection angle, $(\bar{\theta})$. Specific fatigue life denotes the ratio of instantaneous fatigue life (N) to total fatigue life (N_f); the ellipses in (**a**,**b**) show a comparison at a specific point.

For a small crack length, 2a = 100 μm, we obtain η_{175} = 0.96, η_{227} = 0.88 and η_{175}/η_{227} = 1.09. This means that there is still a substantial level of RICC for both stress amplitudes. For a longer crack of 2a = 300 μm, however, the result is η_{175} = 0.64, η_{227} = 0.25 and η_{175}/η_{227} = 2.56. Interestingly, the level of RICC for σ_{a2} (=227 MPa) becomes significantly lower than that for σ_{a1} (=175 MPa). In the case of 2a = 500 μm, it holds η_{175} = 0.26, η_{227} = 0.014, η_{175}/η_{227} = 18.6 and the level of RICC for σ_{a2} (=227 MPa) already becomes negligible. The above comparative analysis clearly shows that there is a lower level of RICC related to the higher applied stress amplitude σ_{a2} (=227 MPa) for all fatigue cracks longer than 100 μm, which corresponds to a higher crack growth rate under the applied stress amplitude σ_{a2} (=227 MPa) than that for σ_{a1} (=175 MPa). This clearly explains the slow crack growth up to around 100 μm of sample under the applied stress amplitude. Crack coalescence under the applied σ_{a2} (=227 MPa) happened for cracks much longer than 100 μm with the level of RICC lower (or even negligible) compared to that under the stress σ_{a1} (=175 MPa). Therefore, crack coalescence is considered to compensate for lower RICC level for σ_{a2} (=227 MPa). On the other hand, the crack coalescence could compensate for the retardation of crack growth rate caused by a longer crack path due to a higher tortuosity (i.e., zig-zag growth) of the crack when σ_{a2} (=227 MPa). This retardation is considered to be directly proportional to the linear roughness ratio, R_{L175}/R_{L227} [21]. Indeed, the kinking geometry does not change the level of the maximum shear stress ahead of the crack front; therefore, the related decrease of K_{Ia} (geometrical shielding) has no considerable effect on the crack growth rate [22].

5. Conclusions

Fatigue crack growth behaviour such as crack propagation and its features, roughness-induced crack closure and fatigue striation with respect to various stress amplitudes (175, 200 and 227 MPa), were studied here for CP Ti. The followings are the main outcomes of the investigation:

- Number of cycles to failure increased from 1.05×10^5 to 3.40×10^5 to 6.94×10^5 with decreasing stress amplitude from 227 to 200 to 175 MPa, respectively. Throughout the total life cycle, most of the cycle was spent on crack growth for all stress conditions. It was found that cracks initiated from the surface of the samples.
- A single fatigue crack was observed to be responsible for failure of sample tested at 175 MPa. At 200 MPa, two main cracks, and at 227 MPa three cracks were recorded for

fatigue failure. Each of the cracks at 227 MPa propagated via sub-crack coalescence. In all conditions, crack deflection, crack branching and slip bands, which are characteristics of crack closure, were noticed. Beyond the initial 100 μm, crack growth rate for the 227 MPa sample was higher than that of the 175 MPa sample.
- RICC calculation for crack lengths of 100, 300 and 500 μm under the 175 and 227 MPa conditions showed a remarkable outcome. Up to 500 μm crack length a substantial RICC was calculated at 175 MPa, while the same level of RICC for 227 MPa was found up to 300 μm. Beyond 100 μm, the RICC level for 175 MPa displayed a higher value than at 227 MPa. This gives a reasonable explanation for the abrupt increase of crack growth rate under the 227 MPa condition.

Author Contributions: Conceptualization, M.A. and D.R.; validation, M.A.; investigation, M.A., M.S.I.; resources, R.C., S.Y.; data curation, M.A.; writing—original draft preparation, M.A., M.S.I.; writing—review and editing, D.R.; visualization, M.A., M.S.I.; supervision, D.R.; funding acquisition, M.A. All authors have read and agreed to the published version of the manuscript.

Funding: This project has received funding from Enterprise Ireland and the European Union's Horizon 2020 Research and Innovation Programme under the Marie Sklodowska-Curie grant agreement No 847402.

Institutional Review Board Statement: Not applicable.

Informed Consent Statement: Not applicable.

Data Availability Statement: Not applicable.

Conflicts of Interest: The authors declare no conflict of interest.

References

1. Takao, K.; Kusukwa, K. Low-cycle fatigue behaviour of commercially pure titanium. *Mater. Sci. Eng. A* **1996**, *213*, 81–85. [CrossRef]
2. Wang, S.-H. A study on the change of fatigue fracture mode in two titanium alloys. *Fatigue Fract. Eng. Mater. Struct.* **1998**, *21*, 1077–1087. [CrossRef]
3. Suresh, S. Crack deflection: Implications for the growth of long and short fatigue cracks. *Metall. Trans. A* **1983**, *14*, 2375–2385. [CrossRef]
4. Ogawa, T.; Tokaji, K.; Ohya, K. The effect of microstructure and fracture surface roughness on fatigue crack propagation in a Ti-6Al-4V alloy. *Fatigue Fract. Eng. Mater. Struct.* **1993**, *16*, 973–982. [CrossRef]
5. Llorca, J. Roughness-induced fatigue crack closure: A numerical study. *Fatigue Fract. Eng. Mater. Struct.* **1992**, *15*, 655–669. [CrossRef]
6. Wang, S.-H.; Müller, C. Fracture surface roughness and roughness-induced fatigue crack closure in Ti-2.5 wt% Cu. *Mater. Sci. Eng. A* **1998**, *255*, 7–15. [CrossRef]
7. Suresh, S. Fatigue crack deflection and fracture surface contact: Micromechanical models. *Metall. Trans. A* **1985**, *16*, 249–260. [CrossRef]
8. Antunes, F.V.; Ramalho, A.; Ferreira, J.M. Identification of fatigue crack propagation modes by means of roughness measurements. *Int. J. Fatigue* **2000**, *22*, 781–788. [CrossRef]
9. Ritchie, R.O.; Suresh, S. Some considerations on fatigue crack closure at near-threshold stress intensities due to fracture surface morphology. *Metall. Trans. A* **1982**, *13*, 937–940. [CrossRef]
10. Ding, Y.S.; Tsay, L.W.; Chen, C. The effects of hydrogen on fatigue crack growth behaviour of Ti–6Al–4V and Ti–4.5Al–3V–2Mo–2Fe alloys. *Corros. Sci.* **2009**, *51*, 1413–1419. [CrossRef]
11. Okayasu, M.; Chen, D.; Wang, Z. Experimental study of the effect of loading condition on fracture surface contact features and crack closure behavior in a carbon steel. *Eng. Fract. Mech.* **2006**, *73*, 1117–1132. [CrossRef]
12. Student, O.Z.; Cichosz, P.; Szymkowski, J. Correlation between the fracture roughness and fatigue threshold of high-temperature degraded steel. *Mater. Sci.* **1999**, *35*, 796–801. [CrossRef]
13. Jiang, X.P.; Wang, X.Y.; Li, J.X.; Li, D.Y.; Man, C.S.; Shepard, M.J.; Zhai, T. Enhancement of fatigue and corrosion properties of pure Ti by sandblasting. *Mater. Sci. Eng. A* **2006**, *429*, 30–35. [CrossRef]
14. Wojcik, C.C.; Chan, K.S.; Koss, D.A. Stage I fatigue crack propagation in a titanium alloy. *Acta Metall.* **1988**, *36*, 1261–1270. [CrossRef]
15. Ismarrubie, Z.N.; Sugano, M. Environmental effects on fatigue failure micromechanisms in titanium. *Mater. Sci. Eng. A* **2004**, *386*, 222–233. [CrossRef]

16. Nikitin, A.; Bathias, C.; Palin-Luc, T.; Shanyavskiy, A. Crack path in aeronautical titanium alloy under ultrasonic torsion loading. *Frat. Integrita Strutt.* **2015**, *10*, 213–222. [CrossRef]
17. Fintová, S.; Arzaghi, M.; Kuběna, I.; Kunz, L.; Sarrazin-Baudoux, C. Fatigue crack propagation in UFG Ti grade 4 processed by severe plastic deformation. *Int. J. Fatigue* **2017**, *98*, 187–194. [CrossRef]
18. Małecka, J.; Rozumek, D. Metallographic and Mechanical Research of the O–Ti_2AlNb Alloy. *Materials* **2020**, *13*, 3006. [CrossRef]
19. Gray, G.T.; Williams, J.C.; Thompson, A.W. Roughness-Induced Crack Closure: An Explanation for Microstructurally Sensitive Fatigue Crack Growth. *Metall. Trans. A* **1983**, *14*, 421–433. [CrossRef]
20. Pokluda, J.; Pippan, R. Analysis of roughness-induced crack closure based on asymmetric crack-wake plasticity and size ratio effect. *Mater. Sci. Eng. A* **2007**, *462*, 355–358. [CrossRef]
21. Pokluda, J.; Sandera, P. *Micromechanisms of Fracture and Fatigue*; Springer: London, UK, 2010.
22. Pippan, R. The crack driving force for fatigue crack propagation. *Eng. Fract. Mech.* **1993**, *44*, 821–829. [CrossRef]

Article

Fatigue Variability of Alloy 625 Thin-Tube Brazed Specimens

Seulbi Lee [1,†], Hanjong Kim [2,†], Seonghun Park [2,*] and Yoon Suk Choi [1,*]

1. School of Materials Science and Engineering, Pusan National University, Busan 46241, Korea; seulbi0921@pusan.ac.kr
2. School of Mechanical Engineering, Pusan National University, Busan 46241, Korea; hkim@pusan.ac.kr
* Correspondence: paks@pusan.ac.kr (S.P.); choiys@pusan.ac.kr (Y.S.C.); Tel.: +82-51-510-2330 (S.P.); +82-51-510-2382 (Y.S.C.); Fax: +82-51-514-0685 (S.P.); +82-51-512-0528 (Y.S.C.)
† All authors contributed equally to this study.

Abstract: As an advanced heat exchanger for aero-turbine applications, a tubular-type heat exchanger was developed. To ensure the optimum performance of the heat exchanger, it is necessary to assess the structural integrity of the tubes, considering the assembly processes such as brazing. In this study, fatigue tests at room temperature and 1000 K were performed for 0.135 mm-thick alloy 625 tubes (outer diameter of 1.5 mm), which were brazed to the grip of the fatigue specimen. The variability in fatigue life was investigated by analyzing the locations of the fatigue failure, fracture surfaces, and microstructures of the brazed joint and tube. At room temperature, the specimens failed near the brazed joint for high σ_{max} values, while both brazed joint failure and tube side failure were observed for low σ_{max} values. The largest variability in fatigue life under the same test conditions was found when one specimen failed in the brazed joint, while the other specimen failed in the middle of the tube. The specimen with brazed joint failure showed multiple crack initiations circumferentially near the surface of the filler metal layer and growth of cracks in the tube, resulting in a short fatigue life. At 1000 K, all the specimens exhibited failure in the middle of the tube. In this case, the short-life specimen showed crack initiation and growth along the grains with large through thickness in addition to multiple crack initiations at the carbides inside the tube. The results suggest that the variability in the fatigue life of the alloy 625 thin-tube brazed specimen is affected by the presence of the brazed joint, as well as the spatial distribution of the grain size and carbides.

Keywords: fatigue variability; alloy 625; thin tube; fractography; microstructure

Citation: Lee, S.; Kim, H.; Park, S.; Choi, Y.S. Fatigue Variability of Alloy 625 Thin-Tube Brazed Specimens. Metals 2021, 11, 1162. https://doi.org/10.3390/met11081162

Academic Editor: Dariusz Rozumek

Received: 25 June 2021
Accepted: 19 July 2021
Published: 22 July 2021

Publisher's Note: MDPI stays neutral with regard to jurisdictional claims in published maps and institutional affiliations.

Copyright: © 2021 by the authors. Licensee MDPI, Basel, Switzerland. This article is an open access article distributed under the terms and conditions of the Creative Commons Attribution (CC BY) license (https://creativecommons.org/licenses/by/4.0/).

1. Introduction

Heat exchangers are one of the key components for environmentally friendly gas-turbine engines with lower emissions and higher specific fuel consumption ratings to meet environmental requirements and airline operation conditions [1–5]. Advanced heat exchangers for aero-turbine applications require compact and complicated shapes to achieve high efficiency and size limitations. Such a design limitation sometimes necessitates the use of submillimeter-scale thin tubes to maximize the heat exchange rate in a limited space. However, the use of such thin tubes requires an additional assembly process called "brazing" to connect the thin tubes to the inlet and outlet of the heat exchanger. Here, the mechanical integrity of thin tubes (including the brazed joint) needs to be thoroughly evaluated to ensure the optimum performance of the heat exchanger. However, it is difficult to assess the thermo-mechanical strengths of thin tubes and brazed joints under fluctuating loads, which simulate actual service conditions [6–14].

In the present study, a thin-tube brazed fatigue specimen was designed to evaluate the fatigue properties of thin tubes including brazed joints at room and elevated (1000 K) temperatures, considering the actual operating conditions of heat exchangers for the aero-turbine engine. Here, solid-solution-strengthened Ni-based alloy 625 was chosen as the thin-tube material. Alloy 625 has been used for a variety of components in the aerospace, aeronautics, marine, chemical, and nuclear industries because of its high-temperature strength, corrosion resistance in a variety of environments, excellent fabricability, and weldability (for tubing) [15–18]. Since alloy 625 components in industry fields were subjected to high temperature operations for a long duration, it is important to secure the mechanical properties at service temperature. V. Shankar et al. [19] extensively investigated the tensile properties at intermediate temperatures with various strain rates. L.M. Suave et al. [20] investigated the high temperature fatigue properties and clarified the thermal aging effect on the mechanical properties. Additionally, evaluation of the mechanical properties, considering the structural integrity, was important to estimate the reliable properties of the component. For the brazed superalloys, low-cycle fatigue [21,22], creep [23,24], and thermal cycling [25] have been studied, but high-cycle fatigue has been conducted in few studies [26,27]. In terms of the structural applications for brazed joints, J. Chen et al. [26,27] performed a high-cycle fatigue test of alloy 625 joints brazed with the Palnicro-36M[TM] filler metal to clarify the effect of single-lap joints under various lap distance-to-thickness ratios.

The alloy 625 thin-tube brazed fatigue specimens used in this study were designed similarly to the final configuration installed in an actual heat exchanger to reliably assess the fatigue properties of the alloy 625 thin tubes and brazed joints. A systematic study was conducted to understand the source of fatigue variability observed in alloy 625 thin-tube specimens. In particular, an effort was made to interpret the observed fatigue variability in terms of the microstructural variability of thin tubes, including the brazed joint, using microstructure analysis and fractography. The thermo-mechanical responses of the thin tube and brazed joint were also investigated to clarify the fracture mechanism.

2. Experiments

An alloy 625 tube with a thickness of 0.135 mm and an outer diameter of 1.5 mm was prepared by repeated drawing and heat treatment processes after initial tubing via tungsten inert gas welding of a 0.2 mm thick strip roll. Figure 1 shows the geometry of the newly designed thin-tube fatigue specimen. As can be seen, the thin tube was connected to the grip (also made of alloy 625) using brazing, in the same way that a tube is brazed to a tube sheet in an actual heat exchanger. This type of fatigue specimen allows us to investigate the fatigue behavior of the thin tube, including the influence of the brazed joint. The filler metal used for brazing was BNi-2, containing boron and silicon as melting point depressants. The chemical compositions of the alloy 625 tube and filler metal are listed in Table 1.

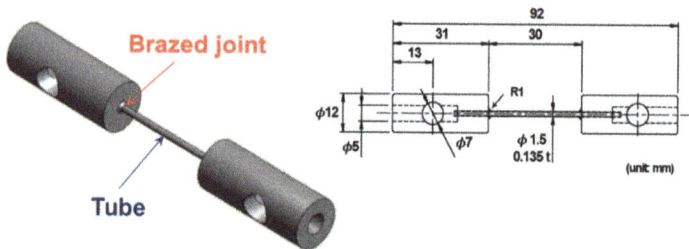

Figure 1. Geometry of the thin-tube brazed fatigue specimen.

Table 1. Chemical compositions of alloy 625 and BNi-2 used in this study.

	Composition (wt.%)											
	Ni	Cr	Fe	Co	Mo	Nb	Ti	Al	C	Mn	Si	B
Alloy 625	61.4	21.2	4.7	0.1	8.6	3.34	0.18	0.14	0.03	0.07	0.20	-
BNi-2	82	7	3	-	-	-	-	-	-	-	4.5	3.1

Before the fatigue tests, the microstructure of the brazed specimen was analyzed. The sample was polished to a 0.04 µM finish using colloidal silica and etched in 15 mL of HCl, 10 mL of acetic acid, and 10 mL of HNO_3. The microstructures were then characterized using optical microscopy and scanning electron microscopy (SEM) equipped with energy dispersive spectroscopy (EDS). To measure the local mechanical properties, the hardness across the brazed joint was measured using a micro-Vickers hardness tester. The applied load was 4.903 N for the dwell time of 10 s.

Fatigue tests were performed using the MTS 810 system (MTS Systems Corp., Eden Prairie, MN, USA). The fatigue specimen shown in Figure 1 was connected to a ball-joint grip with a pin for the tilt- and twist-free alignment of the specimen along the loading direction. The applied load was measured using a 2 kN load cell (BCA-200K, TESTA Corp., Gyeonggi-do, Korea). Fatigue tests were performed under cyclic loading with a stress ratio (R) of 0.1 and a frequency of 10 Hz. Fatigue tests were performed at six different maximum stresses (σ_{max}) for the two temperatures as follows: 495, 526, 557, 587, 618, and 649 MPa for room temperature (RT), and 355, 371, 386, 402, 418, and 433 MPa for 1000 K. Here, three to seven tests were conducted for each test condition. Fatigue test conditions are summarized in Table 2. After the tests, fractography was performed on the failed specimens to clarify the failure mechanisms and fatigue variability of the thin-tube brazed specimens.

Table 2. Fatigue test conditions used in this study.

Test Temperature (K)	Fatigue Limit (Cycles)	Test Type	Stress Ratio, R	Loading Frequency (Hz)	Maximum Stress Level, σ_{max} (MPa)
300	1×10^6	Tension–tension cyclic loading	0.1	10	495, 526, 557, 587, 618, 649
1000					355, 371, 386, 402, 418, 433

3. Results and Discussion

Figure 2 shows the SEM images of the longitudinal and radial cross-sections of an alloy 625 thin tube prior to the fatigue test. The EDS analysis indicated that carbides of (Nb, Ti)C were present in the tube. The size of the carbides distributed outside the tube was larger than those distributed inside the tube. Carbide streaks along the drawing direction were also observed, which originated from the drawing process. Figure 3 shows the metallographs of the radial cross-section of the tube, including the weld zone in Figure 3b. As can be seen, the grain size appears to be finer in the weld zone than in the other regions (base metal). Despite the repeated annealing process after every drawing step, microstructural differences between the weld zone and the base metal are still observed. Therefore, the grain size distribution through the thickness is quite heterogeneous in the range of approximately 1–8 grains.

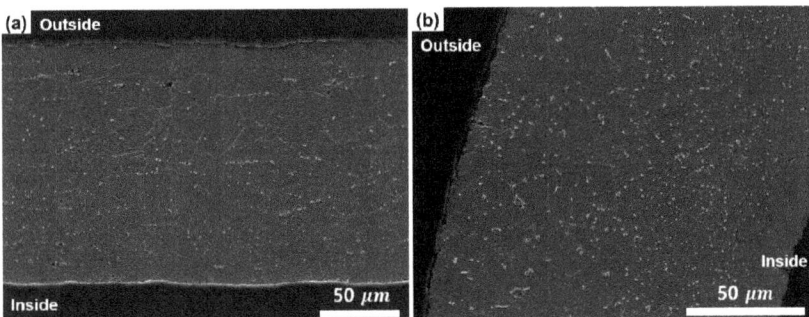

Figure 2. SEM images of an alloy 625 thin tube prior to the fatigue test: (**a**) longitudinal and (**b**) radial cross-sections.

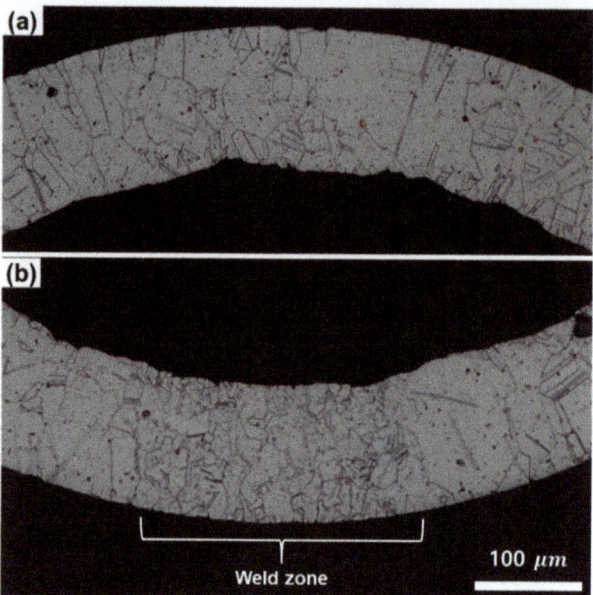

Figure 3. Metallographs at the radial cross-section of the tube: (**a**) typical base metal region and (**b**) region including the weld zone.

Figure 4 shows the SEM images of the cross-section for the as-brazed fatigue specimen. Here, Figure 4b,c presents enlarged images of the areas labeled in Figure 4a. A typical brazing microstructure of Ni-base alloys is apparent, comprising various intermetallic phases [28,29]. An EDS analysis in an athermal solidification zone (Figure 4b) demonstrated that the phases marked by Z1, Z2, and Z3 are nickel boride, Ni–Si–B ternary intermetallic, and eutectic of γ-nickel and fine nickel silicide, respectively. The microstructure at the interface between the filler metal and the base metal shown in Figure 4c consisted of phases marked by Z4 and Z5, which are chromium boride, and the γ-nickel solid solution, respectively. In addition, the microhardness profile across the brazed joint was measured, as displayed in Figure 5. The microhardness in the tube was approximately 182 HV. In the filler metal region, however, the hardness increased to as high as 710 HV. This high hardness value is attributed to intermetallic constituents, including nickel boride, chromium boride, and nickel silicide, which are known for hard and brittle phases.

Fatigue tests were conducted at RT and 1000 K on the thin-tube brazed specimens, and the S–N curves are plotted in Figure 6. Each fatigue test condition shows different variabilities in the fatigue life (N_f). For the RT fatigue tests, the maximum variability in N_f was found at σ_{max} = 495 MPa, which exhibited maximum and minimum N_f values of 2,891,024 and 444,361 cycles, respectively. For the 1000 K fatigue tests, however, an even higher maximum variability in N_f was found at σ_{max} = 402 MPa, which showed maximum and minimum N_f values of 781,877 and 31,909 cycles, respectively. In addition, such a variability in N_f does not seem to show any particular relationship with the level of the maximum stress applied, particularly for the 1000 K fatigue tests. The fatigue strengths determined at N_f = 1 × 10^6 cycles were 511 and 371 MPa at RT and 1000 K, respectively. The modified fatigue strengths with various standard deviations are listed in Table 3.

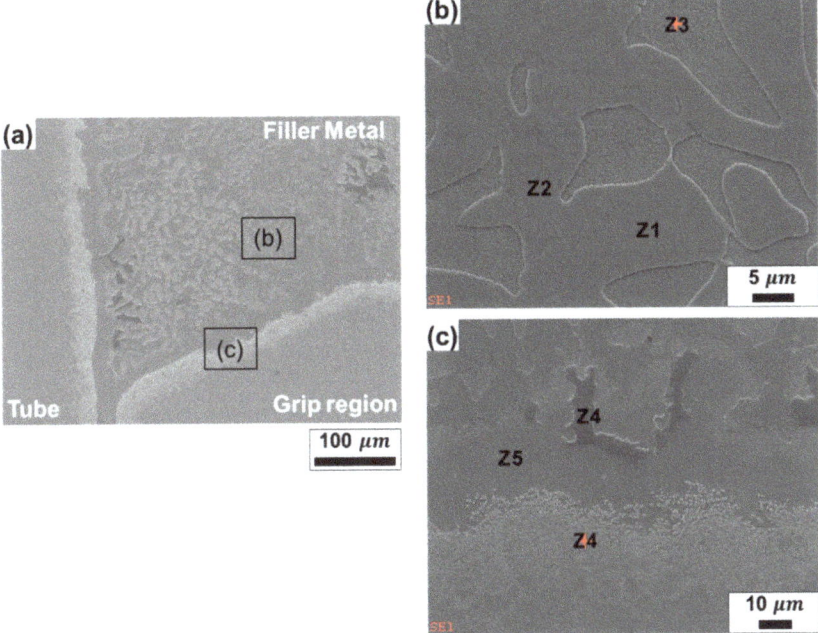

Figure 4. (a) SEM image of the brazed region; (b,c) enlarged SEM images of the areas marked in (a).

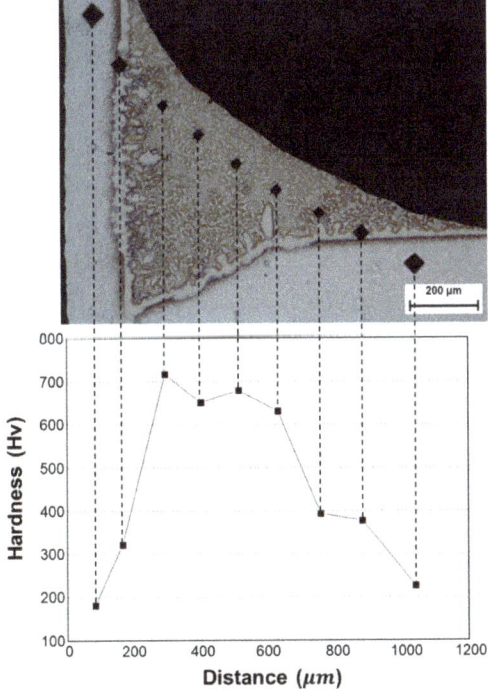

Figure 5. Microhardness profile across the brazed region.

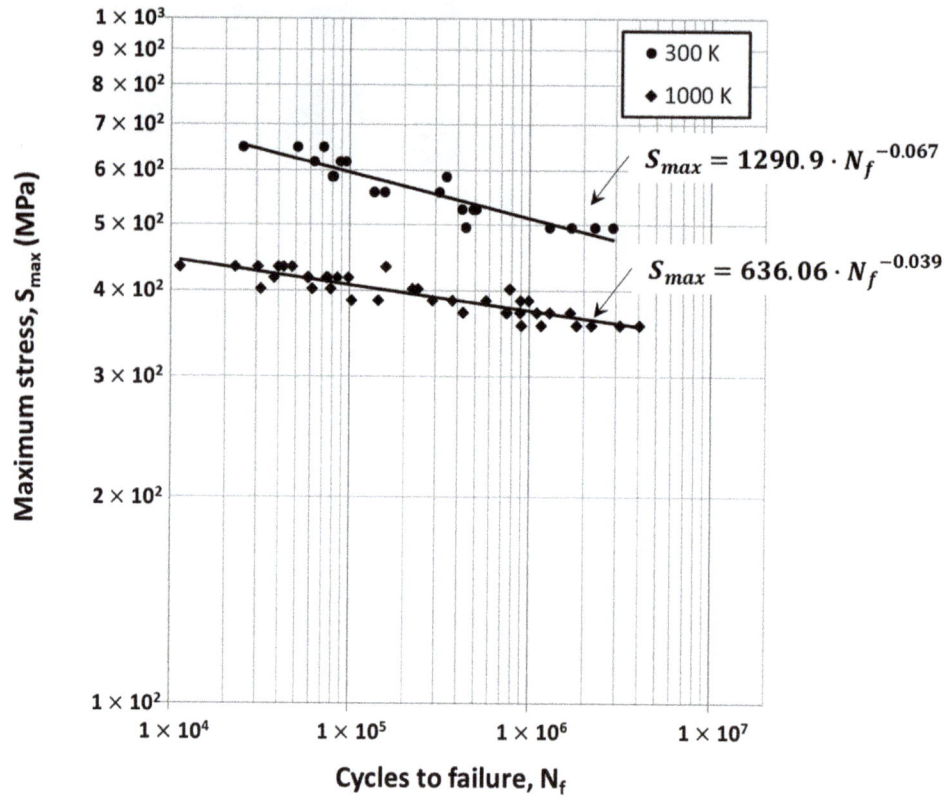

Figure 6. S–N plots for all fatigue data tested at RT and 1000 K.

Table 3. Modified fatigue strength at 10^6 cycles with various standard deviations.

Standard Deviation	Fatigue Strength (MPa)		Reliability (%)
	RT	1000 K	
σ	490	357	84.2
2σ	468	343	97.7
3σ	446	329	99.9

Because the thin-tube fatigue specimen has an unusual geometry, i.e., a thin tube brazed to the grip (Figure 1), further analysis was performed to clarify whether the brazed joint affected the variability of the fatigue life. In Figure 7, the fatigue fracture locations of all tested specimens are plotted as a function of N_f for different applied stresses (σ_{max}). Here, the fracture location (%) is the percentile fracture location relative to the distance from the grip section: 0% for fracturing at the brazed joint and 50% for fracturing in the middle of the thin tube. As can be seen, the fatigue fracture at RT was mainly near the brazed joint for high σ_{max} (approximately $\sigma_{max} \geq 557$ MPa and $N_f < 400{,}000$ cycles), while the fatigue fracture was either near the brazed joint or in the thin tube for low σ_{max} ($N_f > 400{,}000$ cycles). However, the fatigue fracture at 1000 K was mainly in the thin tube away from the brazed joint, regardless of the magnitude of σ_{max}. This result implies that the presence of the brazed joint affects the fatigue life mainly for RT fatigue at a high σ_{max}.

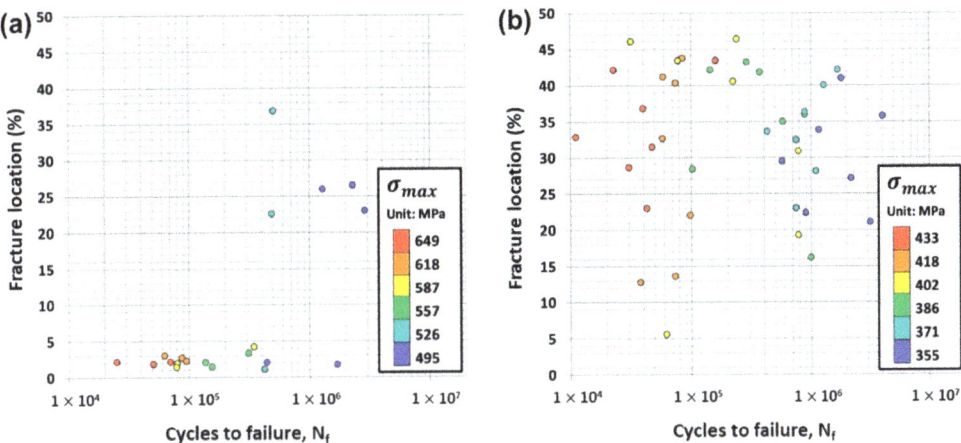

Figure 7. Variations of the fracture location as a function of fatigue life (N_f) for different σ_{max} values at (**a**) RT and (**b**) 1000 K.

To further investigate the relationship between the fracture location and the fatigue life at RT, fractography was performed for the two specimens, which showed a difference of approximately seven times in fatigue life even under the same test conditions (σ_{max} = 495 MPa). Here, the short-life tube specimen fractured near the brazed joint at 444,361 cycles, whereas the long-life tube specimen fractured in the middle of the tube at 2,891,024 cycles, as shown in Figure 8a,b, respectively. Figure 9 shows the fractography of the two specimens tested at RT and σ_{max} = 495 MPa. In Figure 9, the crack initiation sites are indicated by arrows. The fracture surfaces of the short-life specimen presented in Figure 9a,b show typical brittle fractures with multiple crack initiations (indicated by the arrow in Figure 9b) almost everywhere near the surrounding filler metal surfaces. These initial cracks circumferentially formed on the surfaces of the filler metal seem to propagate toward the inside of the thin tube, leading to premature fatigue failure. The gradual change in the fracture type from the transgranular quasi-cleavage fracture in the filler metal layer of the outer tube to the relatively dimpled fracture of the inner tube indirectly supports crack propagation. The microhardness profile displayed in Figure 5 indicates that the filler metal layer in the vicinity of the brazed joint is brittle, as expected, compared with the thin tube, owing to the presence of various intermetallic compounds (see Figure 4). Moreover, a certain degree of stress concentration is expected at the brazed joint owing to the geometric discontinuity (a notch effect) between the thin tube and the grip. The stress concentration factor of the fillet in the brazed joint can be calculated quantitatively using the factor of k_f shown in Equation (1) [30]:

$$k_f = 0.268 \left(\frac{D_0}{r} \right) + 0.998 \qquad (1)$$

where k_f is the stress concentration factor and D_0 and r the outer diameter and radius of fillet, respectively. By Equation (1), the k_f is calculated at 1.40, which is well-coincident with S.H. Kang et al.'s study [31]. They verified the local mechanical response of alloy 625 brazed tubes with BNi-2 filler metal by considering the geometry and the local material properties of the brazed part, using a finite element method. Under these circumstances, the stress caused by cycling loading tends to be unevenly distributed in the brazed joint, and the incompatible deformation response of the filler metal layer (due to the presence of different intermetallic compounds) facilitate crack initiation on the surfaces of the filler metal near the brazed joint, causing a relatively short fatigue life.

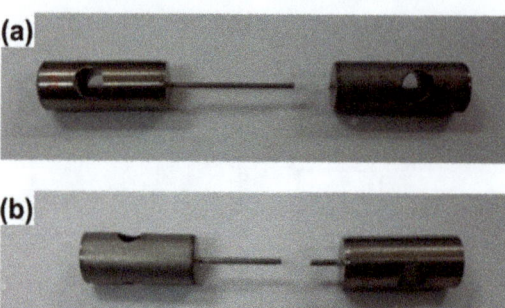

Figure 8. (a) Short-life (444,361 cycles) and (b) long-life (2,891,024 cycles) tube specimens at RT and σ_{max} = 495 MPa.

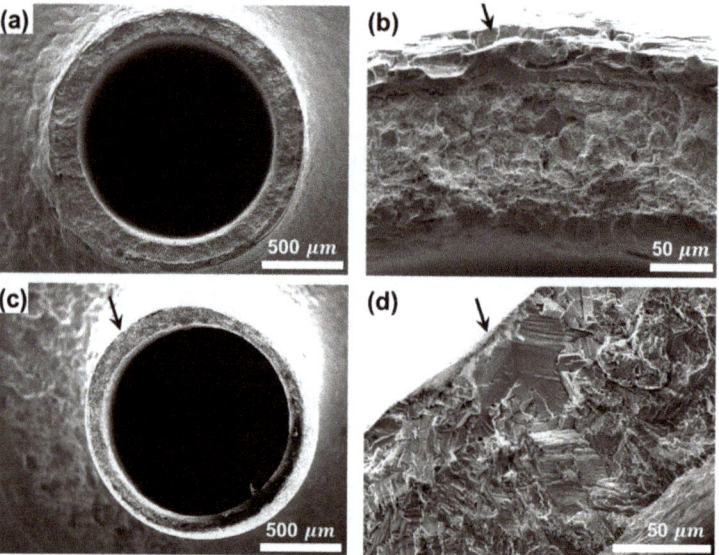

Figure 9. Fracture surfaces for (a,b) short-life (444,361 cycles) and (c,d) long-life (2,891,024 cycles) specimens tested at RT and σ_{max} = 495 MPa.

In contrast to the short-life specimen, the long-life specimen, which fractured in the middle of the tube (Figure 8b), exhibited single crack initiation near the outer surface, as shown in Figure 9c,d. It can be seen that the initial crack formed a facet perpendicular to the loading direction (Figure 9d), which is typical for high-cycle fatigue. Unlike the short-life specimen, which displays simultaneous crack propagation from the filler metal layer of the outer tube to the inner tube, the long-life specimen shows a single crack propagating from the outer tube surface through the thickness, then spreading out to the neighboring area.

The fatigue lives at 1000 K also exhibited large variability, as shown in Figure 6. However, almost all fatigue fractures were observed in the region of the thin tube and not in the brazed joint, as shown in Figure 7b. Accordingly, fractography was performed for the two specimens, which showed a difference of approximately 10 times in fatigue life (N_f = 781,887 and 78,621 cycles for the long-life and short-life specimens, respectively) at 1000 K and σ_{max} = 402 MPa. As shown in Figure 10, both specimens displayed failure in the tube region away from the brazed joint. Figure 11 presents the fracture surfaces for the short- and long-life specimens tested at 1000 K and σ_{max} = 402 MPa. Regarding the short-life specimen, multiple crack initiations both at the outer surface and carbides inside the tube were observed, as indicated by the arrows in Figure 11b,c. In particular, cracks initiated at the outer surface (arrow in Figure 11b) showed progress through the tube thickness. Metallographs (Figure 12a,c) taken directly underneath the fracture surface presented in Figure 11a indicate that the grain size at the crack initiation site (marked with an arrow) is very large (only approximately two grains through the thickness). In addition, based on the non-uniform microstructure distribution compared to the surrounding area, the crack initiation site corresponds to the weld zone. Hence, during cyclic loading at 1000 K, the heterogeneous grain distribution can cause premature fracture, particularly in the weld zone.

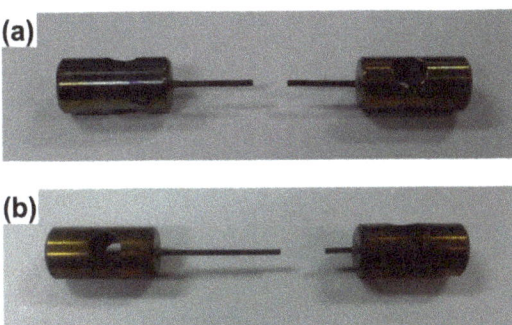

Figure 10. (a) Short-life (78,621 cycles) and (b) long-life (781,887 cycles) tube specimens at 1000 K and σ_{max} = 402 MPa.

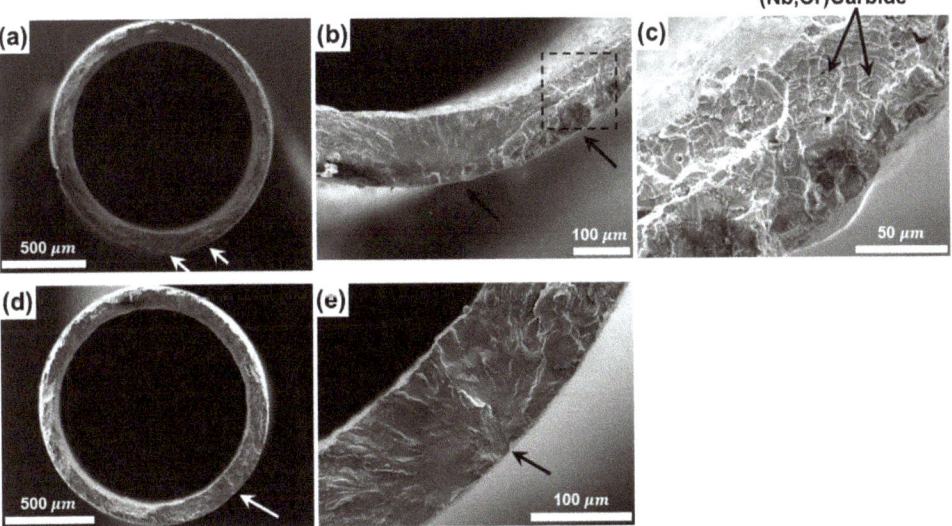

Figure 11. Fracture surfaces for (a–c) short-life (78,621 cycles) and (d,e) long-life (781,887 cycles) specimens tested at 1000 K and σ_{max} = 402 MPa. The arrows in (a,b,d,e) indicate the crack initiation sites. (c) presents the enlarged image of the area highlighted in (b).

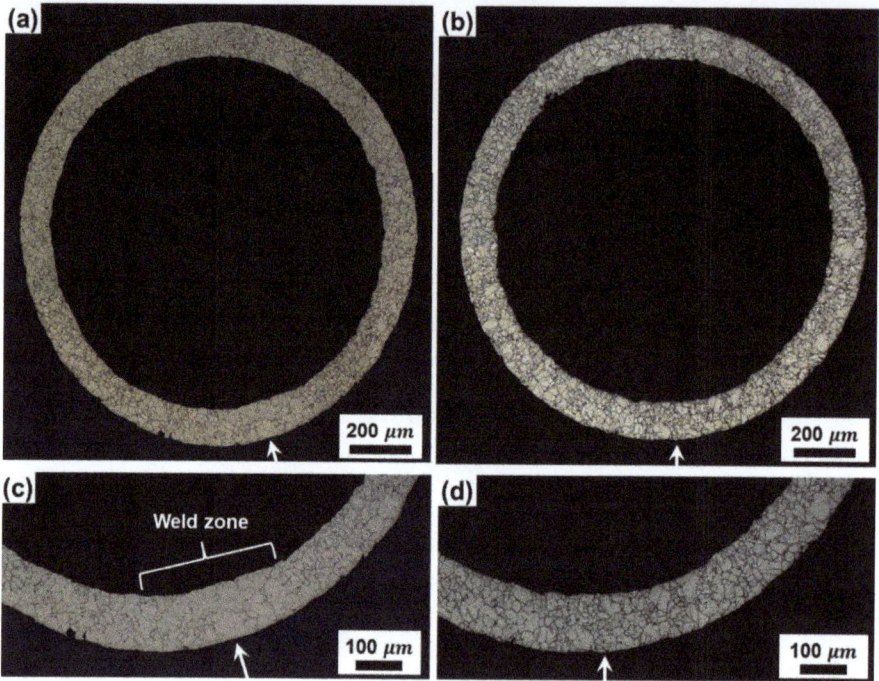

Figure 12. Microstructures directly underneath the fracture surfaces fatigue-tested at 1000 K and σmax = 402 MPa for (**a,c**) short-life (78,621 cycles) and (**b,d**) long-life (781,887 cycles) specimens. The arrows indicate the locations of the fatigue crack initiation.

For the long-life specimen, a single crack initiation was observed near the outer surface, as shown in Figure 11d,e. Here, the initial surface crack is indicated by an arrow in Figure 11e. Metallographs (Figure 12b,d) obtained directly underneath the fracture surface presented in Figure 11d show homogeneous grain distribution over the tube. This indicates that, for the long-life specimen, if the inhomogeneity of the weld zone does not directly lead to fracture at the beginning of the fatigue test, the effect of grain size on the fatigue life is reduced owing to homogenization by long-term exposure at 1000 K.

Combining the results of the fatigue life variability for the alloy 625 thin-tube brazed specimens tested at RT and 1000 K, the following factors were found to affect the fatigue variability: the brazed joint (particularly, the filler metal layer at the joint) and the spatial distribution of the grain size and carbides. The presence of the brazed joint shown in Figure 1 (and the filler metal layer provided in Figures 4 and 5) can cause a notch stress concentration effect. Hence, the filler metal layer in the brazed part can act as a crack initiation site, particularly for the RT fatigue and under high σ_{max}, because the various intermetallic phases in the filler metal layer, as well as the geometrical effect of the brazed part, cause local deformation incompatibility under cycling loading. The fatigue crack initiation in the filler metal layer (near the brazed joint) occurred at high σ_{max} values (approximately ≥ 557 MPa) and resulted in short fatigue lives (N_f < 400,000 cycles), as shown in Figure 7a. In this case, multiple cracks initiated circumferentially in the filler metal layer and propagated inward into the thin tube, as shown in Figure 9a,b. The largest fatigue life variability at RT was found when one specimen failed near the brazed joint, whereas the other specimen failed in the tube region, as shown in Figure 8, at σ_{max} = 495 MPa, which seems to be in a transient stress range between the brazed joint failure and the tube failure (see Figure 7a). This result indicates that the presence of the brazed joint causes variability in fatigue life, particularly for low σ_{max} values.

For 1000 K fatigue, however, no apparent brazed joint failure was observed, as shown in Figure 7b. This is because the deformation incompatibility among different intermetallic phases in the filler metal layer was fully accommodated (even under cycling loading) at such a high temperature [32]. In this case, the spatial distribution of the grain size and carbides seems to affect the fatigue life variability. The presence of a large near-surface grain (corresponding to the weld zone), which has approximately 1–2 through-thickness grains and can facilitate the initiation of a fatigue crack near the outer surface, as shown in Figure 11b,c, leads to a short fatigue life. In particular, the fatigue life will be even shorter if multiple crack initiations at carbides inside the tube occur simultaneously, as shown in Figure 11b,c, in addition to the crack initiation at a large near-surface grain.

4. Conclusions

Variability in fatigue life at room temperature and 1000 K was investigated for brazed alloy 625 thin-tube specimens. The fatigue life variability was found to be influenced by the presence of the brazed joint (and its properties), as well as the spatial distribution of the grain size and carbides.

At room temperature, a correlation between the fracture location and fatigue life was observed. Specimens tested at $\sigma_{max} \geq 557$ MPa exhibited failure near the brazed joint and relatively short fatigue lives (typically, $N_f < 400,000$ cycles). For $\sigma_{max} < 557$ MPa, however, a short-life specimen failed at the brazed joint, whereas a long-life specimen failed in the middle of the tube. The brazed-joint failed specimens showed multiple crack initiations circumferentially in the filler metal layer and growth of cracks through the thickness of the tube, leading to a short fatigue life.

At 1000 K, all test specimens failed in the middle of the tube. Specifically, the short-life specimen showed fatigue crack initiation and growth at a location with only 1–2 through-thickness grains. Crack growth seemed to be further facilitated by multiple crack initiations at the carbides inside the tube. In conclusion, homogeneous grain distribution within the tube and small grains through the tube thickness can prevent premature fracture, leading to a long fatigue life.

Author Contributions: Research conceptualization and design, Y.S.C. and S.P.; evaluation of the mechanical properties, H.K.; microstructure analysis, S.L.; data analysis and writing S.L. and H.K. All authors have read and agreed to the published version of the manuscript.

Funding: This work was supported by the Korea Institute of Energy Technology Evaluation and Planning (KETEP) and the Ministry of Trade, Industry & Energy (MOTIE) of the Republic of Korea (No. 20193310100050) and by a two-year research grant from Pusan National University.

Data Availability Statement: Not applicable.

Conflicts of Interest: The authors declare that they have no conflict of interest.

References

1. McDonald, C.F.; Wilson, D.G. The utilization of recuperated and regenerated engine cycles for high-efficiency gas turbines in the 21st century. *Appl. Therm. Eng.* **1996**, *16*, 635–653. [CrossRef]
2. Wilfert, G.; Sieber, J.; Rolt, A.; Baker, N.; Touyeras, A.; Colantuoni, S. New Environmental Friendly Aero Engine Core Concepts. In Proceedings of the XVIII International Symposium of Air Breathing Engines, The International Society for Air Breathing Engines (ISABE), Beijing, China, 2–7 September 2007.
3. Jeong, J.H.; Kim, L.S.; Lee, J.K.; Ha, M.Y.; Kim, K.S.; Ahn, Y.C. Review of heat exchanger studies for high-efficiency gas turbines. In Proceedings of the Turbo Expo: Power for Land, Sea, and Air, American Society of Mechanical Engineers, Montreal, CA, USA, 14–17 May 2007.
4. McDonald, C.F. Recuperator considerations for future higher efficiency microturbines. *Appl. Therm. Eng.* **2003**, *23*, 1463–1487. [CrossRef]
5. Min, J.K.; Jeong, J.H.; Ha, M.Y.; Kim, K.S. High temperature heat exchanger studies for applications to gas turbines. *Heat Mass Transf.* **2009**, *46*, 175–186. [CrossRef]
6. Bruening, G.B.; Chang, W.S. Cooled cooling air systems for turbine thermal management. In Proceedings of the ASME 1999 International Gas Turbine and Aeroengine Congress and Exhibition, Proceeding of the Turbo Expo: Power for Land, Sea, and Air, American Society of Mechanical Engineers, Indianapolis, IN, USA, 7–10 June 1999.

7. Maziasz, P.J.; Pint, B.A.; Shingledecker, J.P.; Evans, N.D.; Yamamoto, Y.; More, K.L.; Lara-Curzio, E. Advanced alloys for compact, high-efficiency, high-temperature heat-exchangers. *Int. J. Hydrog. Energ.* **2007**, *32*, 3622–3630. [CrossRef]
8. MacKenzie, P.M.; Walker, C.A.; McKelvie, J. A method for evaluating the mechanical performance of thin-walled titanium tubes. *Thin-Walled Struct.* **2007**, *45*, 400–406. [CrossRef]
9. Sonsino, C.M.; Bruder, T.; Baumgartner, J. S-N Lines for welded thin joints-suggested slopes and fat values for applying the notch stress concept with various reference radii. *Weld. J.* **2010**, *54*, R375–R392. [CrossRef]
10. Fayard, J.L.; Bignonnet, A.; Van, K.D. Fatigue design criterion for welded structures. *Fatigue Fract. Eng. Mater. Struct.* **1996**, *19*, 723–729. [CrossRef]
11. Lazzarin, P.; Sonsino, C.M.; Zambardi, R. A notch stress intensity approach to assess the multiaxial fatigue strength of welded tube-to-flange joints subjected to combined loadings. *Fatigue Fract. Eng. Mater. Struct.* **2004**, *27*, 127–140. [CrossRef]
12. Yokobori, T.; Yamanouchi, H.; Yamamoto, S. Low cycle fatigue of thin-walled hollow cylindrical specimens of mild steel in uni-axial and torsional tests at constant strain amplitude. *Int. J. Fract.* **1965**, *1*, 3–13. [CrossRef]
13. Carstensen, J.; Mayer, H.; Brøndsted, P. Very high cycle regime fatigue of thin walled tubes made from austenitic stainless steel. *Fatigue Fract. Eng. Mater. Struct.* **2002**, *25*, 837–844. [CrossRef]
14. Baumgartner, J.; Tillmann, W.; Bobzin, K.; Ote, M.; Wiesner, S.; Sievers, N. Fatigue of brazed joints made of X5CrNi18-10 and Cu110 and derivation of reliable assessment approaches. *Weld. World.* **2020**, *64*, 707–719. [CrossRef]
15. Floreen, S.; Fuchs, G.E.; Yang, W.J. The Metallurgy of Alloy 625. *Superalloys* **1994**, *718*, 13–37.
16. Radavich, J.F.; Fort, A. Effects of Long-Time Exposure in Alloy 625 at 1200 °F, 1400 °F and 1600 °F. *Superalloys* **1994**, *718*, 635–647.
17. Conder, C.R.; Smith, G.D. Microstructural and mechanical property characterization of aged Inconel625LCF. In *Proceedings of the International Symposium on Superalloys 718, 625, 706 and Various Derivatives*; The Minerals, Metals, and Materials Society (TMS): Pittsburgh, PA, USA, 1997; pp. 447–458.
18. Marchese, G.; Beretta, M.; Aversa, A.; Biamino, S. In situ alloying of a modified Inconel 625 via laser powder bed fusion: Microstructure and mechanical properties. *Metals* **2021**, *11*, 988. [CrossRef]
19. Shankar, V.; Valsan, M.; Rao, K.B.S.; Mannan, S.L. Effects of temperature and strain rate on tensile properties and activation energy for dynamic strain aging in alloy 625. *Metall. Mater. Trans. A: Phys. Metall. Mater. Sci.* **2004**, *35A*, 3129–3139. [CrossRef]
20. Suave, L.M.; Cormier, J.; Bertheau, D.; Villechaise, R.; Soula, A.; Hervier, Z.; Hamon, F. High temperature low cycle fatigue properties of alloy 625. *Mater. Sci. Eng. A* **2016**, *630*, 161–170. [CrossRef]
21. Osoba, L.O.; Ojo, O.A. Influence of solid-state diffusion during equilibration on microstructure and fatigue life of superalloy wide gap brazements. *Metall. Mater. Trans. A* **2013**, *44*, 4020–4024. [CrossRef]
22. Yang, X.; Dong, C.; Shi, D.; Zhang, L. Experimental Investigation on Both Low Cycle Fatigue and Fracture Behavior of DZ125 Base Metal and the Brazed Joint at Elevated Temperature. *Mater. Sci. Eng. A* **2011**, *528*, 7005–7011. [CrossRef]
23. Shi, D.; Dong, C.; Yang, X.; Sun, Y.; Wang, J.; Liu, J. Creep and fatigue lifetime analysis of directionally solidified superalloy and its brazed joints based on continuum damage mechanics at elevated temperature. *Mater. Des.* **2013**, *45*, 643–652. [CrossRef]
24. Kim, Y.H.; Kim, I.H.; Kim, C.S. *Asian Pacific Conference for Fracture and Strength*, 4th ed.; Trans Tech Publications Ltd.: Bäch SZ, Switzerland, 2005; pp. 2876–2882.
25. Kim, Y.H.; Kim, K.T. *Multifunctional Materials and Structures*; Trans Tech Publications Ltd.: Bäch SZ, Switzerland, 2008; pp. 894–897.
26. Chen, J.; Deners, V.; Turner, D.P.; Bocher, P. Experimental investigation on high-cycle fatigue of Inconel 625 superalloy brazed joints. *Metall. Mater. Trans. A: Phys. Metall. Mater. Sci.* **2018**, *49A*, 1244–1253. [CrossRef]
27. Chen, J.; Demers, V.; Cadotte, E.-L.; Turner, D.; Bocher, P. Structural performance of Inconel 625 superalloy brazed joints. *J. Mater. Eng. Perform.* **2017**, *26*, 547–553. [CrossRef]
28. Pouranvari, M.; Ekrami, A.; Kokabi, A.H. Solidification and solid state phenomena during TLP bonding of IN718 superalloy using Ni–Si–B ternary filler alloy. *J. Alloys Compd.* **2013**, *563*, 143–149. [CrossRef]
29. Tung, S.K.; Lim, L.C.; Lai, M.O. Solidification phenomena in nickel base brazes containing boron and silicon. *Scr. Mater.* **1996**, *34*, 763–769. [CrossRef]
30. Garrell, M.G.; Shih, A.J.; Lara-Curzio, E.; Scattergood, R.O. Finite-element analysis of stress concentration in ASTM D 638 tension specimens. *J. Test. Eval.* **2003**, *31*, 1–6.
31. Kang, S.-H.; Park, S.H.; Min, J.K.; Cho, J.R.; Ha, M.-Y. Evaluation of mechanical integrity on the brazing joint of a tube-type heat exchanger with considering local material properties. *Proc. Inst. Mech. Eng. C* **2013**, *227*, 420–433. [CrossRef]
32. Zhang, Y.-C.; Yu, X.-T.; Jiang, W.; Tu, S.-T.; Zhang, X.-C. Elastic modulus and hardness characterization for microregion of Inconel 625/BNi-2 vacuum brazed joint by high temperature nanoindentation. *Vacuum* **2020**, *181*, 109582. [CrossRef]

Article

Fatigue Behavior of Nonreinforced Hand-Holes in Aluminum Light Poles

Cameron R. Rusnak [1],* and Craig C. Menzemer [2]

[1] Research Assistant, M.ASCE, Auburn Science and Engineering Center (ASEC 210), Department of Civil Engineering, The University of Akron, Akron, OH 44325, USA
[2] Professor, Auburn Science and Engineering Center (ASEC 210), Department of Civil Engineering, The University of Akron, Akron, OH 44325, USA; ccmenze@uakron.edu
* Correspondence: crr44@uakron.edu

Abstract: Hand-holes are present within the body of welded aluminum light poles. They are used to provide access to the electrical wiring for both installation and maintenance purposes. Wind is the main loading on these slender aluminum light poles and acts in a very cyclic way. In the field, localized fatigue cracking has been observed. This includes areas around hand-holes, most of which are reinforced with a cast insert welded to the pole. This study is focused on an alternative design, specifically hand-holes without reinforcement. Nine poles with 18 openings were fatigue tested in four-point bending at various stress ranges. Among the 18 hand-holes tested, 17 failed in one way or another as a result of fatigue cracking. Typically, fatigue cracking would occur at either the 3:00 or 9:00 positions around the hand-hole and then proceed to propagate transversely into the pole before failure. Finite element analysis was used to complement the experimental study. Models were created with varying aspect ratios to see if the hand-hole geometry had an effect on fatigue life.

Keywords: aluminum hand-hole; nonreinforced hand-hole; fatigue test; design S-N curve; high cycle fatigue

Citation: Rusnak, C.R.; Menzemer, C.C. Fatigue Behavior of Nonreinforced Hand-Holes in Aluminum Light Poles. *Metals* **2021**, *11*, 1222. https://doi.org/10.3390/met11081222

Academic Editor: Dariusz Rozumek

Received: 6 July 2021
Accepted: 28 July 2021
Published: 30 July 2021

Publisher's Note: MDPI stays neutral with regard to jurisdictional claims in published maps and institutional affiliations.

Copyright: © 2021 by the authors. Licensee MDPI, Basel, Switzerland. This article is an open access article distributed under the terms and conditions of the Creative Commons Attribution (CC BY) license (https://creativecommons.org/licenses/by/4.0/).

1. Introduction

Aluminum light poles support overhead light fixtures and are used to are illuminate sidewalks, roadways, parking lots, recreational areas, and others. This is due to its light weight, resistance to corrosion, high strength to weight ratio, and ease of handling and joining. Wind is the main contribution to loading to these light poles, which can be classified as slender structures. Fatigue cracking can occur in either steel or aluminum when exposed to any kind of repeated loads. In these aluminum light poles, electrical wires will run through conduit, into the hollow section of the pole, and then proceed up into the light [1,2].

Stress concentrations occur when there are changes within the cross-sectional area of a structural member, examples of which include connections, copes, keyways, cutouts, and others. In modern fatigue design, specifications will use the lower bound S-N curve established from full-scale test data as it will be the first to fail [3,4]. A series of S-N curves represents a ranking of the stress concentration condition that is associated with different mechanical and structural details. One way to increase fatigue life would be to minimize or eliminate abrupt changes in the cross-section or provide smooth, gradual transitions. In aluminum light poles, there are multiple structural details of interest. These include the base to pole connection, mono-tube or truss arm joints truss, and the hand-holes used for electrical access. The behavior of the fatigue in these electrical access hand-holes within these aluminum light poles is largely unknown. The majority of the existing data that have been collected were from large, welded steel poles [5].

Report number 176 from NCHRP (National cooperative Highway Research Program) web only found results of unreinforced and reinforced hand-hole fatigue tests for welded steel structures. Lehigh University studied detail associated with steel light poles under fatigue. During these experiments, 13 of the specimens had handholes with different geometries (as compared with aluminum poles). During testing, zero of the hand-holes failed or cracked. A finite element study was conducted and was used to provide an estimate of how the stress concentration around the pole and hand-hole. On the basis of the analysis, the research found the fatigue resistance of both the unreinforced and reinforced hand-holes to align with AASHTO (American Association of State High and Transportation Officals) Category E of the design S-N curves [6].

Field observations have shown that some hand-hole details are susceptible to failure due to fatigue cracking. NCHRP report number 469 [5] described these fatigue cracks of welded steel structures near the hand-hole in multiple states. These included California, New Mexico, New York, and Minnesota. Subsequent inspections of different poles occurred after the failure of a high mast light pole in Iowa. It was found that another tower had crack associated with the hand-hole. Cracks were found in some of the aluminum light poles along the Mullica River Bridge after a violent storm in 2011 [7]. A picture of one of these failures was taken and can be seen in Figure 1.

Figure 1. Aluminum light pole containing a fatigue crack within the field.

At the University of Akron, a study was conducted on 20 light-pole specimens, with fatigue tests conducted under bending loads. In addition to fatigue test, several static ones were conducted in order to see how the strain distributed around the hand-hole. This study found that the data from the welded hand-hole fatigue tests fell above the category D and E design S-N curves of the Aluminum Design Manual [8]. Another study found that the change in diameter of the pole has a modest effect on the fatigue life. In this companion study, seven eight-inch poles with 14 details were tested [9].

Nine aluminum light poles, each containing two separate hand-holes, were tested in fatigue under four-point bending. All poles were supplied to the University of Akron and were manufactured to standards typical for the industry. Finite element models were created to help improve the understanding of the stress concentrated around the hand-holes. The models created had different aspect ratios.

2. Materials and Methods

2.1. Pole Geometry and Material Properties

Nine aluminum light poles were tested under cyclic loading to examine the behavior of the unreinforced hand-hole. These specimens consisted of a 10 in (25.4 cm) diameter aluminum alloy extruded tube with a 1/4 in (0.635 cm) thick wall. Each of the tubes was fabricated from aluminum alloy 6063. Typically, there is a reinforcement welded into the hand-hole opening, but these tests used only open holes (Figure 2). Hand-holes measured 6 in (150 mm) along the length of the pole in the longitudinal direction of the pole and 4 in (100 mm) in the transverse direction. Each specimen was 144 in (3.66 m), or 12 feet in length, with the hand-holes placed 54 in (1.37 m) in from either end. Support rollers for the specimens were inserted 6 in (15.2 cm) from each end [10].

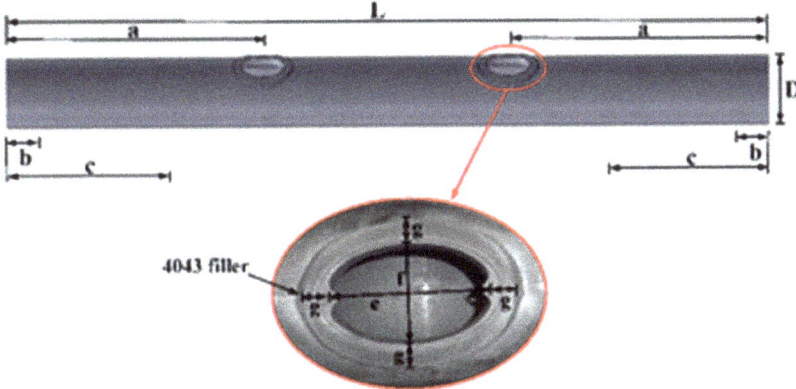

Figure 2. Typical geometry of a welded aluminum hand-hole detail in four-point fatigue testing. (In this study there was no welded detail). (L = 3.66 m (144 in); D = 25.4 cm (10 in); a = 1.37 m (54 in); b = 15.2 cm (6 in); e = 150 mm (6 in); f = 100 mm (4 in)).

Table 1 is a summary of the minimum mechanical properties of the aluminum tube.

Table 1. Mechanical properties of the aluminum tube.

Part Name	Alloy	Tensile Yield Strength	Ultimate Tensile Strength
Tube	6063-T6	213.7 MPa (31 ksi)	241.3 MPa (35 ksi)

2.2. Fatigue Tests

Figure 3 depicts a photo of the four-point bending fatigue test setup in the lab. During testing, a 55 kips (245 KN) MTS servo-hydraulic actuator (MTS Systems Headquarters, Eden Prairie, MN 55344, USA) along with a control system was used to apply the loads to the specimens. The actuator itself was mounted to a load frame capable of safely supporting 300 kips (1335 KN). Loads were applied to each of the specimens through a spreader beam that was attached to the hydraulic system. The supports the specimens rested on consisted of rollers that were machined to fit the cylindrical profile of the specimens.

Figure 3. Four point bending fatigue test set-up.

Testing was conducted with a load control, while the strains were monitored using gages that were applied around the hand-holes, along with a single gage placed in the middle of the specimen. The typical location of the strain gages can be seen in Figure 4. Strain gages had a resistance of 350 ohms and were 1/8 in (3.175 mm) in length. Strain readings were taken every two hours intermittently for 10 seconds using a Micro-Measurements System 8000 data acquisition device (Micro-Measurements A VPG Brand Raleigh, NC 27611, USA) that was wired to the strain gages.

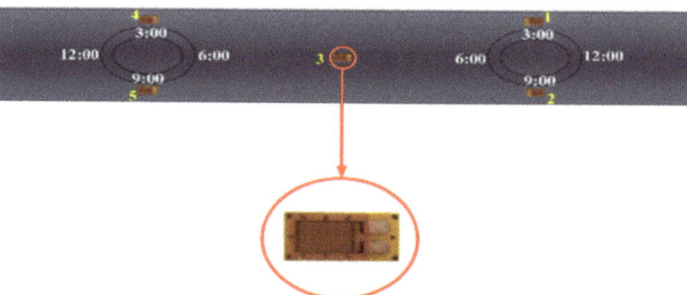

Figure 4. Typical strain gage location and position around the hand-holes.

All of the specimens were oriented with the hand-holes facing downwards direction so that they were in tension during cyclic testing. Failure was achieved when the hand-hole region was cracked to the point that the loading on the specimen could no longer be supported. A maximum displacement was placed on the specimens for each of the fatigue tests to ensure that both hand-holes could be tested. The upper limit was set 10% larger than the maximum static displacement. When this maximum displacement was exceeded, the test would automatically shut down. Once this limit was reached and the test was stopped, the detail that failed was reinforced with a moment clamp. An image half of the moment clamp can be seen in Figure 5. Two halves were placed around the handhole and mechanically secured. The test was restarted and continued until the other hand-hole failed. With the specimen being loaded in four-point bending, the moment and hence applied stress does not change at the undamaged handhole after the moment clamp was applied. Of the nine specimens tested, only one resulted in a catastrophic failure where repair of the detail was not possible.

Figure 5. Half of the moment clamp used to reinforce the hand-hole after initial failure.

Nine poles, each with two hand-holes, were tested at stress ranges between 17.24 MPa (2.5 ksi) and 58.61 MPa (8.5 ksi). Figure 6 depicts a sketch of where the strain gages were installed adjacent to the hand-hole. Strain gages were placed at the 3 and 9 o'clock positions, with the addition of a strain gage in the middle of the specimen. This gage was within 2 to 3 times the tube thickness away from the edge of each hand-hole. All of the strain gages were wired to the data acquisition system to measure the applied strains. Five specimens were cycled at 1 Hz and four were cycled at 2 Hz. Testing continued around the clock. Visual inspections of the hand-holes were conducted daily.

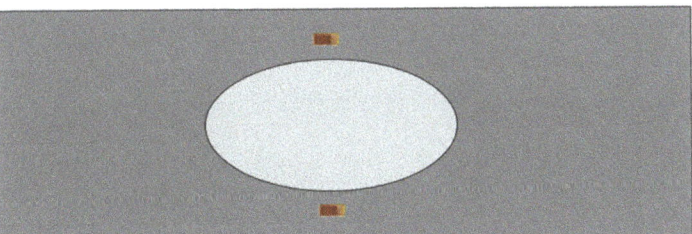

Figure 6. The position of strain gages installed around a hand-hole.

2.3. Finite Element Models

The finite element (FE) model was created for the four-point bending specimens in an attempt to gain a better understanding of the stress distribution around and adjacent to the hand-holes. All modeling was completed within Abaqus CAE (2018 version, Troy, MI 48084, USA). The material model was a general, linear elastic material with only the modulus of elasticity and Poisson's ratio specified. The mechanical response and the influence of geometry on local stresses was the primary concern and focus of the FE analysis. As such, an advanced material model was not needed. Models were classified as "shell" models. Loading was applied by selecting the outermost nodes on the transverse plane where the rollers made contact with the pole. In the model, 365 individual nodes were selected, with a concentrated load of 0.01926 N of force applied to each. A total of 7.03 N was applied to the tube where the rollers were located. Multiple models were created with different aspect ratios. These included 2-1, 1.75-1, 1.5-1, 1-1, 1-1.25, and 1-1.5. The aspect ratio was calculated by dividing length of the hand-hole by the transverse dimension of the handhole. The purpose of the analyses was to determine whether a change in the aspect ratio had any effect on the local stress distribution around the hand-hole. Local stresses were mesh-dependent for this study. A finer mesh size typically increases the stresses local to important geometric details, whereas a more course mesh often results in a reduction in local stresses. The elements in the model consisted of a mix of both hexahedral and tetrahedral element types. Figure 7 depicts the mesh of the 1.5-1 model. In the field, the hand-holes with reinforcement have an aspect ratio of 1.5-1.

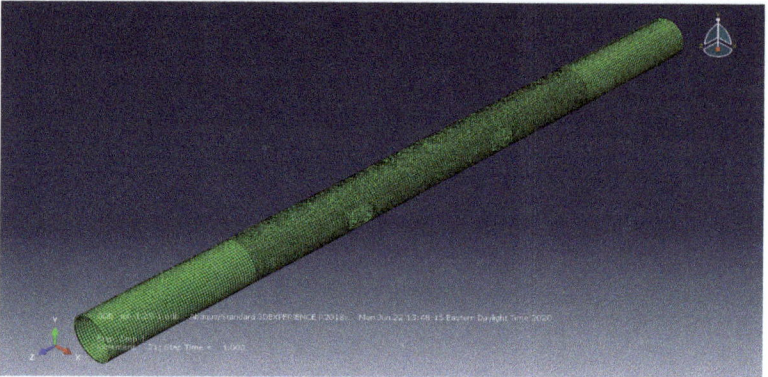

Figure 7. Overall FE model of 1.5-1.

3. Results

3.1. Fatigue Tests Results

Of the 18 hand-holes tested, 17 failed, with the results shown in Figure 8.

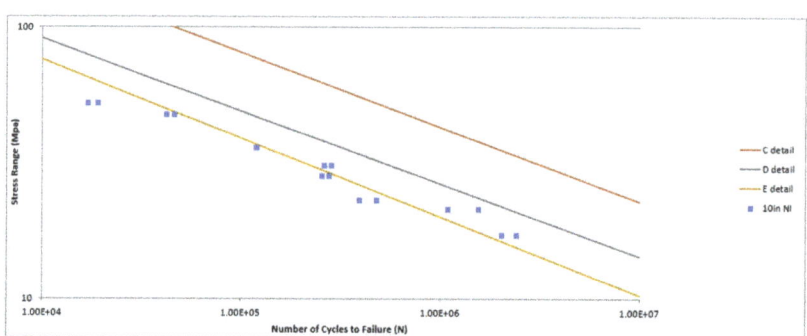

Figure 8. Fatigue test results.

All of the fatigue test data appears to follow the lower bound "E detail" S-N curve, even though the handholes themselves were not reinforced. Additional data would be needed to establish a lower bound for the unreinforced handholes but would be expected to be lower than Category E. Figure 9 shows a comparison between no insert specimens and a previous study conducted on reinforced hand-holes in 10 in diameter poles. This figure clearly demonstrates the benefit of having the welded reinforcement around the hand-holes [11].

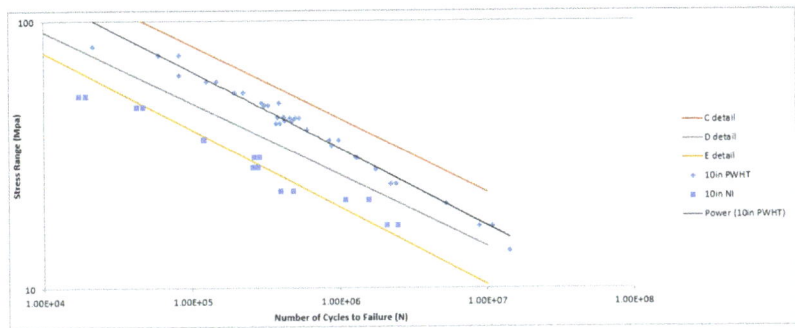

Figure 9. No insert detail vs. old 10 in data.

During fatigue testing, cracks were observed at either the 3:00 or 9:00 position along the minor axis of the hand-hole. Typically, these cracks would initiate and then propagate transversely into the pole from the point of origin. Cracks would become visible and quickly progress into the pole, followed by failure. Figure 10 depicts a fatigue crack at the 3:00 position.

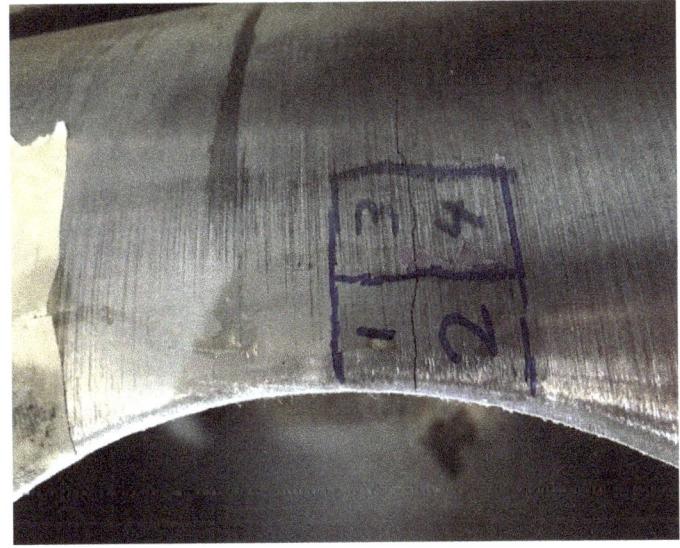

Figure 10. Fatigue crack through pole.

3.2. Finite Element Results

"Hot spots" are generally as local areas with elevated stresses and provide an indication where fatigue cracks may develop. Figures 11–16 depict the stress contour maps along the Z-axis (longitudinal stress) for handholes with different aspect ratios. In all cases, hot spots were most prevalent at either the 3:00 or 9:00 position. The stress concentrations make sense due to how the loading is applied and how the hand-holes themselves are simply an opening within the specimen. The maximum stress increased as the aspect ratio changed from 2–1 to 1–1.5.

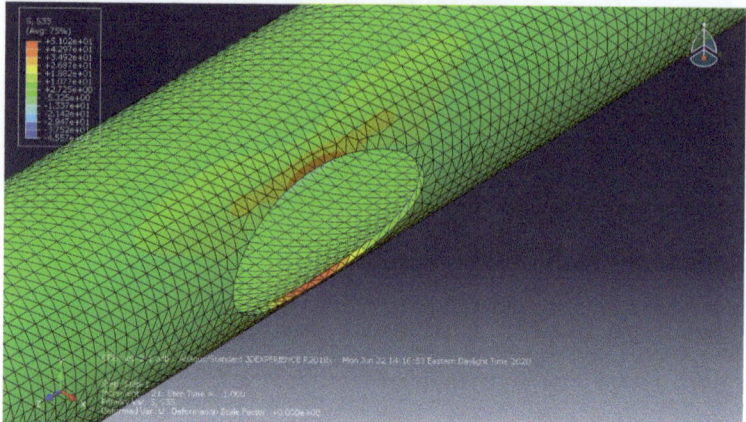

Figure 11. Aspect ratio 2-1 in longitudinal (Z) direction.

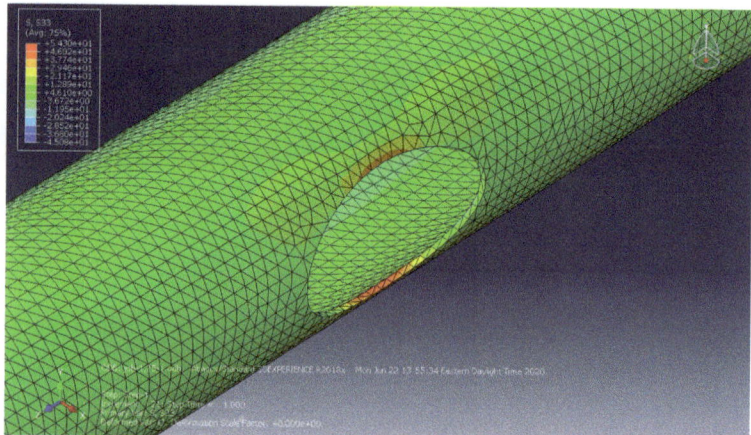

Figure 12. Aspect ratio 1.5-1 in longitudinal (Z) direction.

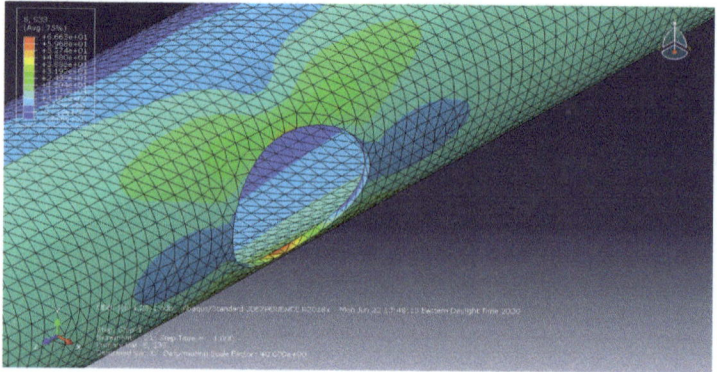

Figure 13. Aspect ratio 1.25-1 in longitudinal (Z) direction.

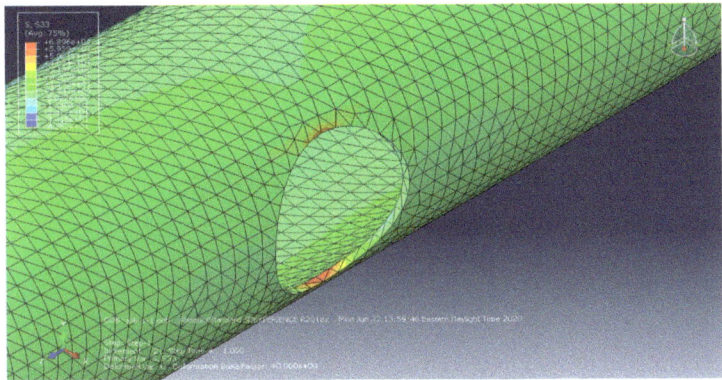

Figure 14. Aspect ratio 1-1 in longitudinal (Z) direction.

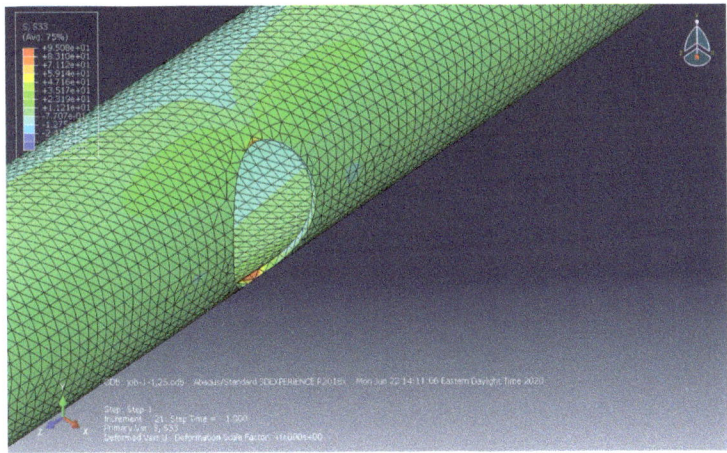

Figure 15. Aspect ratio 1-1.25 in longitudinal (Z) direction.

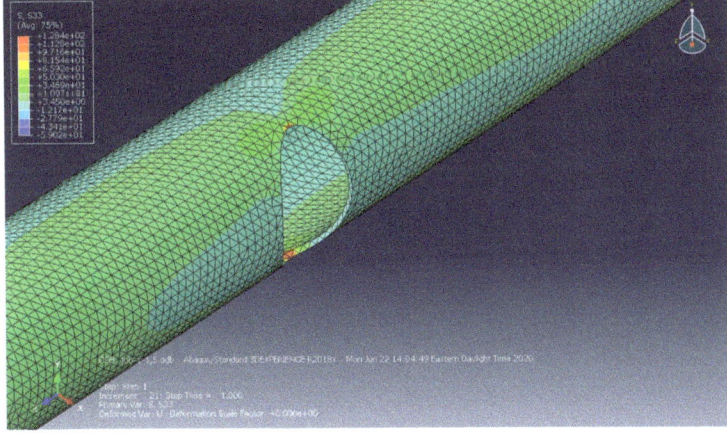

Figure 16. Aspect ratio 1-1.5 in longitudinal (Z) direction.

Figures 17–22 show the maximum transverse stresses along the X-axis (transverse stress) around the hand-holes with different aspect ratios. The transverse stresses follow the same pattern as the longitudinal, with the 1-1.5 aspect ratio having the largest local stresses and the 2-1 having the mildest. These figures also show how stress accumulates inside of the pole apart from the hand-hole opening itself.

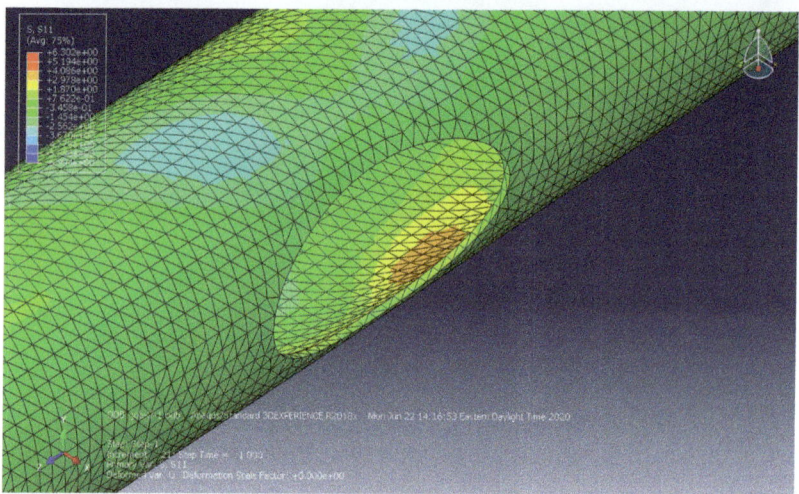

Figure 17. Aspect ratio 2-1 in transverse (X) direction.

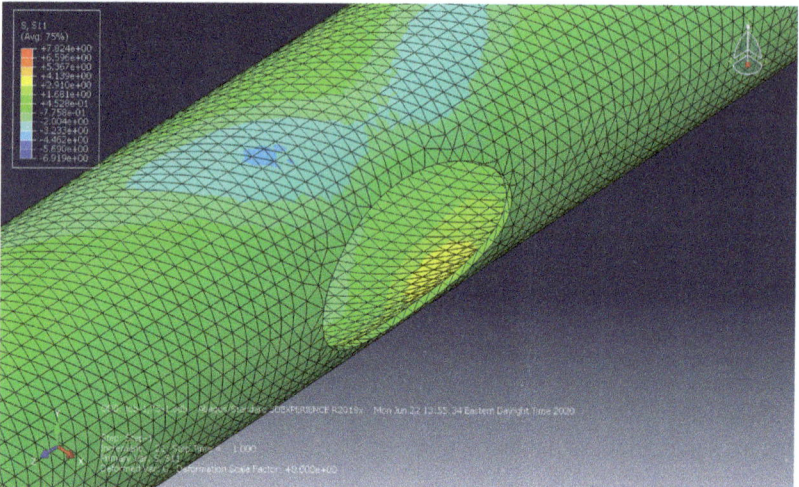

Figure 18. Aspect ratio 1.5-1 in transverse (X) direction.

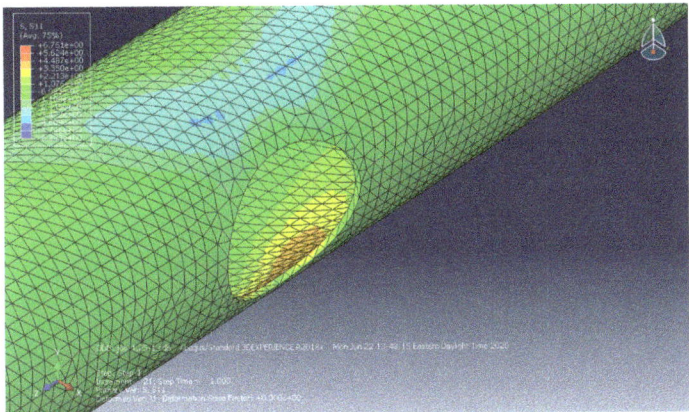

Figure 19. Aspect ratio 1.25-1 in transverse (X) direction.

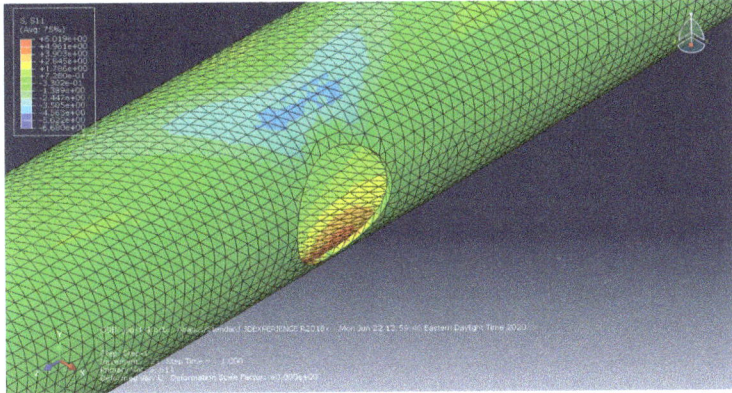

Figure 20. Aspect ratio 1-1 in transverse (X) direction.

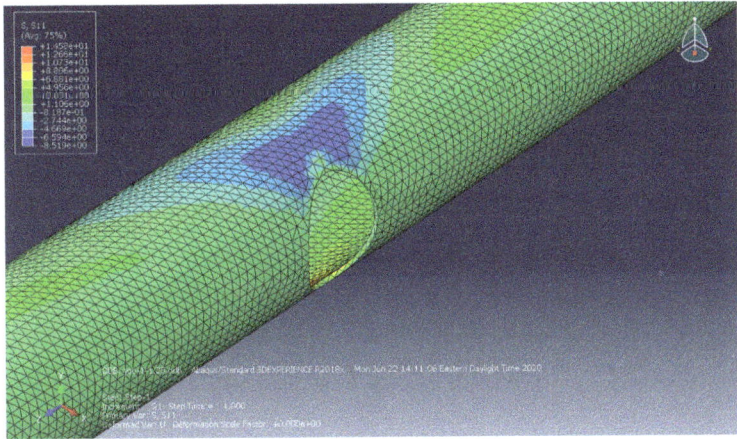

Figure 21. Aspect ratio 1-1.25 in transverse (X) direction.

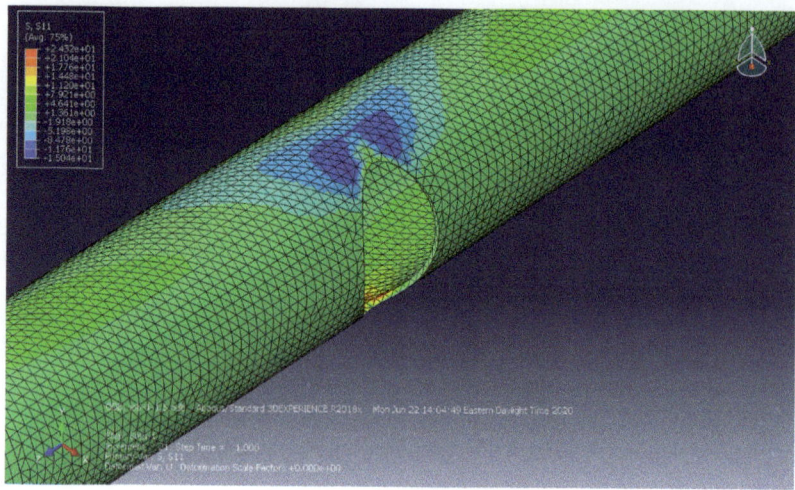

Figure 22. Aspect ratio 1-1.5 in transverse (X) direction.

The maximum longitudinal stress was plotted against the aspect ratio in order to gain a better understanding of how the aspect ratio affects the local stress. The area was taken along the inside of the hand-hole at the 3:00 position. This location was chosen as this spot contained some of the largest stresses (Figure 23). Table 2 accompanies Figure 23 and provides not only longitudinal stresses, but transvers as well. Table 2 shows that the transverse stress was negligible at the location of interest, near the 3:00 position.

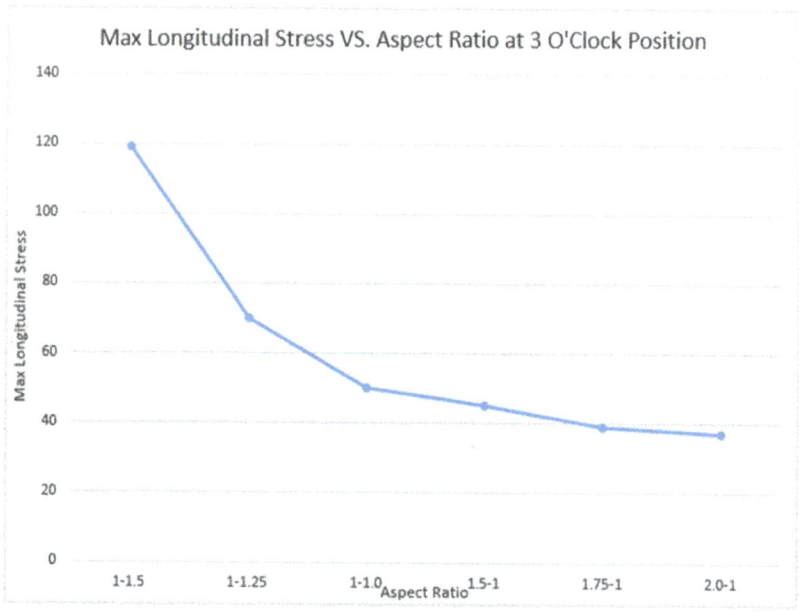

Figure 23. Max longitudinal stress vs. aspect ratio.

Table 2. Longitudinal and transverse stress at the 3 o'clock position for the different aspect ratios.

Aspect Ratio	Longitudinal	Transverses
1-1.5	118.8	5.0
1-1.25	72.0	0.9
1-1	48.4	0.1
1.5-1	44.0	0.0
1.75-1	38.5	0.0
2-1	36.7	0.1

4. Conclusions

Four-point bending fatigue tests were conducted on aluminum light pole containing no reinforcement. Tests revealed that there was a negative effect on the fatigue life when the cast reinforcement was removed. This was most evident when the stress range was higher. This can be seen from Figure 9 from Section 3. A total of nine tests were conducted, resulting in 17 data points, stressed at various different stress ranges. Finite element models showed how different hand-hole aspect ratios affect the stress concentration. While a hand-hole aspect ratio of 2-1 may provide the lowest local longitudinal stress, it may not be the most practical out in the field. It is unknown how a hand-hole with an aspect ratio of 2-1 would behave if reinforced.

Author Contributions: For the duration of, C.R.R. did the conceptualization, methodology, software utilization, validation, formal analysis, investigation, resources, data curation, writing of the original draft, visualization with C.C.M. writing in the form of reviewing and editing. All authors have read and agreed to the published version of the manuscript.

Funding: This research was funded by HAPCO Light Pole Products.

Data Availability Statement: Not applicable.

Conflicts of Interest: The authors declare no conflict of interest.

References

1. Murthy, M.V.V.; Rao, K.P.; Rao, A.K. On stresses around an arbitrarily oriented crack in a cylindrical shell. *Int. J. Struct.* **1974**, *10*, 1243–1269. [CrossRef]
2. Durelli, A.J.; Parks, V.J.; Feng, H.C. Stresses around an elliptical hole in a finite plate subjected to axial loading. *J. Appl. Mech.* **1966**, *33*, 192–195. [CrossRef]
3. The Aluminum Association. *Aluminum Design Manual: Specification for Aluminum Structures*; The Aluminum Association: Arlington, VA, USA, 2010.
4. Fisher, J.W.; Kulak, G.L.; Smith, I.F.C. *A Fatigue Primer for Structural Engineers*; National Steel Bridge Alliance & AISC: Chicago, IL, USA, 1998.
5. Roy, S.; Park, Y.C.; Sause, R.; Fisher, J.W.; Kaufmann, E.J. Cost-effective connection details for highway sign, luminaire, and traffic signal structures. In *NCHRP 10-70 Web-Only Doc. 176*; Transportation Research Board: Washington, DC, USA, 2011.
6. AASHTO (American Association of State Highway and Transportation Officials). *Standard Specifications for Structural Supports for Highway Signs, Luminaires and Traffic Signals (LRFDLTS-1)*; AASHTO (American Association of State Highway and Transportation Officials): Washington, DC, USA, 2015.
7. Dexter, R.J.; Ricker, N.J. Fatigue-Resistant Design of Cantilevered Signal, Sign, and Light Supports. In *NCHRP Report 469*; University of Minnesota: Minneapolis, MN, USA, 2002.
8. Menzemer, C. *Examination of Several Mullica River Bridge Light Poles, Corresponding to J*; Bowman, Hapco: Abingdon, VA, USA, 2012.
9. Daneshkhah, A.R.; Schlatter, C.R.; Rusnak, C.R.; Menzemer, C.C. Fatigue behavior of reinforced welded hand-holes in aluminum light poles. *Eng. Struct. Mater.* **2019**, *188*, 60–68. [CrossRef]
10. Rusnak, C.R. Fatigue Behavior in Reinforced Electrical Access Holes in Aluminum Light Pole Support Structures. Master's Thesis, The University of Akron, Akron, OH, USA, 2019.
11. Hilty, E.; Menzemer, C.; Manigandan, K.; Srivatsan, T. Influence of welding and heat treatment on microstructure, properties and fracture behavior of a wrought aluminum alloy. *Emerg. Mater. Res.* **2014**, *3*, 230–242. [CrossRef]

Article

Application of an Artificial Neural Network to Develop Fracture Toughness Predictor of Ferritic Steels Based on Tensile Test Results

Kenichi Ishihara [1,*], Hayato Kitagawa [1], Yoichi Takagishi [1] and Toshiyuki Meshii [2]

[1] Computational Science Department, KOBELCO RESEARCH INSTITUTE, INC., 1-5-5 Takatsukadai, Nishi-ku, Kobe-shi, Hyogo 651-2271, Japan; kitagawa.hayato@kki.kobelco.com (H.K.); takagishi.yoichi@kki.kobelco.com (Y.T.)

[2] Faculty of Engineering, University of Fukui, 3-9-1 Bunkyo, Fukui-shi, Fukui 910-8507, Japan; meshii@u-fukui.ac.jp

* Correspondence: ishihara.kenichi@kki.kobelco.com; Tel.: +81-78-992-6059

Citation: Ishihara, K.; Kitagawa, H.; Takagishi, Y.; Meshii, T. Application of an Artificial Neural Network to Develop Fracture Toughness Predictor of Ferritic Steels Based on Tensile Test Results. *Metals* **2021**, *11*, 1740. https://doi.org/10.3390/met11111740

Academic Editor: Dariusz Rozumek

Received: 14 October 2021
Accepted: 26 October 2021
Published: 30 October 2021

Publisher's Note: MDPI stays neutral with regard to jurisdictional claims in published maps and institutional affiliations.

Copyright: © 2021 by the authors. Licensee MDPI, Basel, Switzerland. This article is an open access article distributed under the terms and conditions of the Creative Commons Attribution (CC BY) license (https://creativecommons.org/licenses/by/4.0/).

Abstract: Analyzing the structural integrity of ferritic steel structures subjected to large temperature variations requires the collection of the fracture toughness (K_{Jc}) of ferritic steels in the ductile-to-brittle transition region. Consequently, predicting K_{Jc} from minimal testing has been of interest for a long time. In this study, a Windows-ready K_{Jc} predictor based on tensile properties (specifically, yield stress σ_{YSRT} and tensile strength σ_{BRT} at room temperature (RT) and σ_{YS} at K_{Jc} prediction temperature) was developed by applying an artificial neural network (ANN) to 531 K_{Jc} data points. If the σ_{YS} temperature dependence can be adequately described using the Zerilli–Armstrong σ_{YS} master curve (MC), the necessary data for K_{Jc} prediction are reduced to σ_{YSRT} and σ_{BRT}. The developed K_{Jc} predictor successfully predicted K_{Jc} under arbitrary conditions. Compared with the existing ASTM E1921 K_{Jc} MC, the developed K_{Jc} predictor was especially effective in cases where σ_B/σ_{YS} of the material was larger than that of RPV steel.

Keywords: fracture toughness; machine learning; artificial neural network; predictor; yield stress; tensile strength; specimen size

1. Introduction

Both researchers and practitioners have characterized the fracture toughness (K_{Jc}) of ferritic steels in the ductile-to-brittle transition (DBT) region, which is key for analyzing the structural integrity of cracked structures subjected to large temperature changes. K_{Jc} is associated with (I) a large temperature dependence (a change of approximately 400% corresponding to a temperature change of 100 °C) [1–10]; (II) specimen-thickness dependence (roughly, $K_{Jc} \propto 1/(\text{specimen thickness})^{1/4}$) [8,11–21]; and (III) large scatter (approximately ±100% variation around the median value) [8,22,23]. Thus, understanding these three effects is necessary for efficient K_{Jc} data collection.

Since Ritchie and Knott introduced the idea of using critical stress and distance to predict fracture toughness temperature dependence [4], researchers who explicitly or implicitly applied this idea have obtained results that demonstrate a strong correlation between the temperature dependence of fracture toughness and that of yield stress (σ_{YS}) [5,6]. Wallin observed that the increase in fracture toughness with increasing temperature is not sensitive to steel alloying, heat treatment, or irradiation [7]. This observation led to the concept of a universal curve shape that applies to all ferritic steels, i.e., the difference in materials is reflected by the temperature shift. This concept is now known as the master curve (MC) method, as described by the American Society for Testing and Materials (ASTM) E1921 [8]. The existence of a K_{Jc} MC was physically supported by Kirk et al. based on dislocation mechanics considerations [9,10]. They argued that the temperature dependence of K_{Jc} is related to the temperature dependence of the strain energy density (SED). Furthermore, because all steels with body-centered cubic (BCC) lattice structures exhibit a unified σ_{YS} temperature dependence, as described by the Zerilli–Armstrong (Z–A) constitutive model (i.e., Z–A σ_{YS} MC) [24], the existence of a BCC iron lattice structure is the sole factor needed to ensure that K_{Jc} in the DBT region has an MC. Note that Kirk et al. implicitly assumed that the tensile-to-yield stress ratio does not vary with materials, which is not true, and will be a source of deviation from the MC. For example, the failure of this MC to evaluate increases in K_{Jc} at high temperatures has been reported for non-reactor pressure vessel (RPV) steels [25,26]. Despite the successful application of K_{Jc} MC to RPV steels, a reexamination of the basis of K_{Jc} MC existence and additional application limits must be reexamined for the application of ASTM E1921 MC to ferritic steels in general and not be limited to RPV steels.

The size dependence of K_{Jc} has been understood based on the weakest link theory deduced as $K_{Jc} \propto 1/(\text{specimen thickness})^{1/4}$ [17], but because this relationship cannot describe the existence of a lower-bound K_{Jc} for large specimens, researchers have begun to investigate the size dependence of K_{Jc} as the critical stress distribution ahead of a crack-tip requires a second parameter in addition to J (J-A, J-T approach, etc.) [18,19], which is categorized as a crack-tip constraint issue. Consequently, it appears that the development of a deterministic and data-driven size effect formula is possible. ASTM E1921 provides a semi-empirical size effect formula based on the K_{Jc} of a 1-inch-thick specimen, which considers a lower-bound K_{Jc} of 20 MPa·m$^{1/2}$ and proportionality to $1/(\text{specimen thickness})^{1/4}$. There are various opinions regarding this lower-bound value [27–30]; thus, the establishment of a data-driven size effect formula that does not depend on the $\propto 1/(\text{specimen thickness})^{1/4}$ relationship seems possible and necessary.

The statistical nature of fracture toughness has been modeled using the Weibull distribution; some researchers used stress [22] and some used K_{Jc} [8] as the model mean parameter. The idea of using Weibull distributions stems from the understanding that the cleavage fracture can be modeled using the weakest link theory. ASTM E1921 [8] applies a three-parameter Weibull distribution, which assumes a shape parameter of four and a position parameter of 20 MPa·m$^{1/2}$. The failure of this model to predict the scatter in K_{Jc} has also been reported; Weibull parameters (shape and position) vary as functions of the specimen size and temperature, and the parameters differ from those specified in ASTM E1921 [31,32]. If the observed model parameters differ from the assumed parameters, the predicted K_{Jc} and scatter deviate from the measured values. Hence, a more practical method that can potentially prevent the mismatch of the assumed statistical model, i.e., a data-driven approach, is necessary.

Considering the three aforementioned issues, it was considered that a data-driven K_{Jc} predictor that captured features of a variety of BCC metals could improve K_{Jc} prediction accuracy. Another idea was to replace time- and material-consuming fracture toughness tests with tensile tests, assuming that K_{Jc} has a direct relationship with SED obtained via tensile tests. Thus, the artificial neural network (ANN) approach was applied to 531 K_{Jc} data collected in our previous works [30,33] to construct a K_{Jc} predictor based on tensile test properties, thereby eliminating the need to conduct fracture toughness tests. The data were obtained for five heats of RPV and seven heats of non-RPV steels. The widths W of the specimens ranged from 20 to 203.2 mm, and the thickness-to-width ratio B/W was limited to 0.5 (i.e., data obtained with PCCV specimens of $B/W = 1$ were excluded). As a result, a Windows-ready K_{Jc} predictor, which enables K_{Jc} prediction by giving specimen size, tensile and yield stress, was developed. Time- and material-consuming fracture toughness tests are no more necessary.

2. Materials and Methods

2.1. Selection of Machine Learning Model

Machine learning models are used in many fields, such as search engines, image classification, and voice recognition, and various methods have been proposed according to the application. In this study, a tool to predict the fracture toughness K_{Jc} of a material under arbitrary conditions such as the specimen size and temperature, without performing the fracture toughness test, was conducted; this is treated as a regression issue. There are various algorithms for machine learning models for regression. In this study, a multi-layer perceptron (MLP) was classified into an ANN that can express complex nonlinear relationships. The regression model was constructed using the MLP regressor, which is a scikit-learn library of the general-purpose programming language Python [34].

2.2. Overview of Multilayer Perceptron in an Artificial Neural Network

Figure 1 shows a schematic diagram of the MLP network. The MLP is a hierarchical network comprising an input layer, a hidden layer, and an output layer; the unit of the hidden layer is completely connected to the input and output layers [34,35].

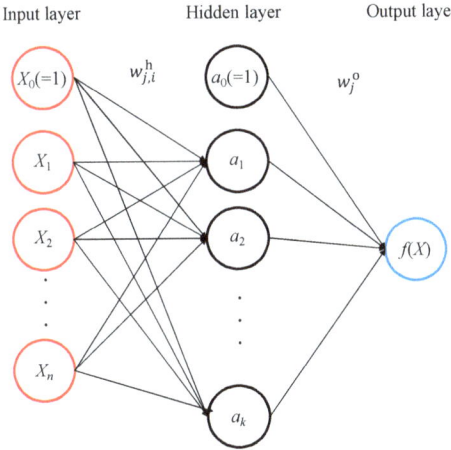

Figure 1. Schematic diagram of multilayer perceptron in an ANN [35].

In Figure 1, only one hidden layer is schematically shown; however, in general, multiple hidden layers are used to enhance the expressiveness of the model. The unit in the hidden layer (hereinafter, referred to as the activation unit a_j ($j = 1 \sim k$)) is calculated using Equation (1), where n input values are X_i and the output values are $f(X)$.

$$a_j = \phi\left(\sum_{i=0}^{n} w_{j,i}^{h} X_i\right) \quad (1)$$

Here, $w_{j,i}^{h}$ is the connection weight, X_0 is a constant called bias, and ϕ of Equation (1) is a function called the activation function. For the activation function, a function with differentiable nonlinearity was selected to enhance the expressiveness of the model. In this study, the rectified linear unit (ReLU) function $\phi(z) = \max(0, z)$ was used and a_j was assigned to the hidden layer. The total number k of a_j (the number of nodes in the hidden layer) and the number of hidden layers are parameters that were adjusted according to the learning accuracy. The output value $f(X)$ can be obtained via Equation (2).

$$f(X) = \phi\left(\sum_{j=0}^{k} w_{j}^{o} a_j\right), \quad (2)$$

where w_{j}^{o} denotes the connection weight. In Equations (1) and (2), the connection weights $w_{j,i}^{h}$, w_{j}^{o} are unknown constants and can be obtained from the combination of known input and output values. By assuming that the known teaching data (true value) are Y

(to distinguish it from $f(X)$, predicted from the input value X_i from Equation (2)), the connection weights can be updated in Equation (3), using the loss function E.

$$E = \frac{1}{2}\sum_{l}(Y_l - f_l(X))^2 + \frac{\alpha}{2}\sum_{l}|w_l^o|^2 \tag{3}$$

Here, the first term in Equation (3) is the sum of the squared residuals of the teaching data Y and the output value $f(X)$, and the second term is a regularization term using the L^2 norm to suppress overfitting. α is a parameter that is adjusted according to learning accuracy. Overfitting is a problem in which training data are overfitted and unknown data cannot be effectively generalized. Several effective optimization algorithms have been developed to avoid falling into a locally optimal solution for updating the connection weights. In this study, adaptive moment estimation (Adam) [36] was used. The connection weight w is updated using Equations (4)–(9).

$$W^{(t)} = w^{(t-1)} - \eta \frac{\hat{m}^{(t)}}{\sqrt{\hat{v}^{(t)}} + \epsilon} \tag{4}$$

$$\hat{m}^{(t)} = \frac{m^{(t)}}{1 - \beta_1^t} \tag{5}$$

$$\hat{v}^{(t)} = \frac{v^{(t)}}{1 - \beta_2^t} \tag{6}$$

$$m^{(t)} = \beta_1 m^{(t-1)} + (1 - \beta_1)\frac{\partial E}{\partial w} \tag{7}$$

$$v^{(t)} = \beta_2 v^{(t-1)} + (1 - \beta_2)\left(\frac{\partial E}{\partial w}\right)^2 \tag{8}$$

$$m^{(0)} = v^{(0)} = 0 \tag{9}$$

The recommended values were used for the adjustment parameters η, β_1, β_2, and ϵ [36]. The error backpropagation method to update the connection weight was used, which calculates the gradient of the loss function by moving backward from the output layer. This method is known to be less computationally expensive than updating weights in the forward direction [37].

2.3. Goodness Valuation of Constructed Learning Model

The goodness of valuation of the constructed machine learning model is based on the coefficient of determination R^2 in Equation (10), where n is the amount of teaching data, Y_i is the true objective value, $f(X)$ is the predicted objective value, and the average value of the true objective values is σ_Y.

$$R^2 = 1 - \frac{\sum_i (Y_i - f_i(X))^2}{\sum_i (Y_i - \mu_Y)^2} \tag{10}$$

The coefficient of determination indicates the goodness of fit of the regression model and is an evaluation index for assessing how well the predicted and true values match. $R^2 = 1$ when the true and predicted values are the same. There is no clear standard for the coefficient of determination, but it can be considered compatible if it is approximately 0.5 or more.

2.4. Dataset

For machine learning, the fracture toughness test data of 531 ferritic steels in the DBTT region obtained by the authors or previous studies were used. Table 1 presents the chemical compositions of the test specimens of the materials considered in the teaching data.

Table 1. Chemical compositions of the test specimens (wt %) of the considered materials.

Heat No.	Material	C	Si	Mn	P	S	Ni	Cr	Mo	V	Cu	Nb	Ti	Al
1	MiuraSFVQ1A [38]	0.18	0.18	1.46	0.002	<0.001	0.90	0.12	0.52	<0.01	-	-	-	-
		0.17	0.17	1.39	0.002	<0.001	0.87	0.11	0.50	<0.01	-	-	-	-
2	Gopalan20MnMoNi55 [39]	0.20	0.24	1.38	0.011	0.005	0.52	0.06	0.30	-	-	0.032	-	0.068
3	ShorehamA533B [40]	0.21	0.24	1.23	0.004	0.008	0.63	0.09	0.53	-	0.08	-	-	0.04
4	MiuraSQV2Ah1 [38]	0.22	0.25	1.44	0.021	0.028	0.54	0.08	0.48	-	0.10	-	-	-
5	MiuraSQV2Ah2 [38]	0.22	0.25	1.46	0.002	0.002	0.69	0.11	0.57	-	-	-	-	-
6	GarciaS275JR [41]	0.18	0.26	1.18	0.012	0.009	<0.085	<0.018	<0.12	<0.02	0.06	-	0.022	0.034
7	GarciaS355J2 [41]	0.2	0.31	1.39	<0.012	0.008	0.09	0.05	<0.12	0.02	0.06	-	0.022	0.014
8	CiceroS460M [42]	0.12	0.45	1.49	0.012	0.001	0.016	0.062	-	0.066	0.011	0.036	0.003	0.048
9	CiceroS690Q [42]	0.15	0.40	1.42	0.006	0.001	0.160	0.020	-	0.058	0.010	0.029	0.003	0.056
10	MeshiiFY2017SCM440 [25]	0.39	0.17	0.62	0.011	0.002	0.07	1.02	0.17	-	0.10	-	-	-
11	MeshiiFY2012S55C [6]	0.55	0.17	0.61	0.015	0.004	0.07	0.08	-	-	0.13	-	-	-
12	MeshiiFY2016S55C [26]	0.54	0.17	0.61	0.014	0.003	0.06	0.12	-	-	-	-	-	-

Tables 2–4 summarize the material heats (heat No. 1–12) used in this study, nT indicates the specimen thickness, and n is expressed in multiples of 25 mm. They are fundamentally extracted from previous work [30,33], but differ slightly in terms of the following: (1) $K_{Jc} > K_{Jc(\text{ulimit})}$ invalid data were excluded, (2) K_{Jc} data were limited to cases obtained with standard specimens of thickness-to-width ratio $B/W = 0.5$, (3) When there were no σ_{YS} data for the fracture toughness test temperature, it was obtained by using the following modified Z–A σ_{YS} temperature-dependent MC [9]

$$\sigma_{0ZA}(T) = \sigma_{0RT} + C_1 exp\left[(T + 273.15)\left(-C_3 + C_4 \log(\dot{\varepsilon})\right)\right] - 49.6 \text{ (MPa)}, \quad (11)$$

where T is the temperature (°C), C_1 = 1033 (MPa), C_3 = 0.00698 (1/K), C_4 = 0.000415 (1/K), and $\dot{\varepsilon}$ = 0.0004 (1/s). The three Miura heats (heat No. 1, 4, 5) were another exception for which linear interpolation of raw data was used because the fracture toughness and tensile test temperatures were different.

Table 2. K_{Jc} data used to construct the proposed tensile property-based MC: RPV steel ASTM A508 equivalent.

Heat No.	Material	Specimen Type	Temps. (°C)	Num. of Temps.	σ_{YS} (MPa)	σ_{YSRT} (MPa)	σ_{BRT} (MPa)	Num. of Specimens	T_0 (°C)
1	MiuraSFVQ1A [38]	1TC(T)	−120~−60	4	530~640	454	594	32	−98
		2TC(T)	−120~−60	4	530~640	454	594	16	−98
		4TC(T)	−100~−80	2	560~607	454	594	12	−98
		0.4TC(T)	−140~−80	4	560~695	454	594	34	−98
		0.4TSE(B)	−140~−80	4	560~695	454	594	29	−98
2	Gopalan20MnMoNi55 [39]	1TC(T)	−140~−80	3	560~667	479	616	18	−133
		0.5TC(T)	−140~−80	3	560~667	479	616	12	−133

Table 3. K_{Jc} data used to construct the proposed tensile property-based K_{Jc} MC: RPV steel ASTM A533B and equivalent.

Heat No.	Material	Specimen Type	Temps. (°C)	Num. of Temps.	σ_{YS} (MPa)	σ_{YSRT} (MPa)	σ_{BRT} (MPa)	Num. of Specimens	T_0 (°C)
3	ShorehamA533B [40]	1TC(T) *	−100~−64	3	551~586	488	644	18	−91
4	MiuraSQV2Ah1 [38]	1TC(T)	−100~−60	3	544~600	473	625	14	−93
		2TC(T)	−100~−60	3	544~600	473	625	14	−93
		4TC(T)	−80~−60	2	544~566	473	625	12	−93
		0.4TC(T)	−120~−60	4	544~658	473	625	32	−93
		0.4TSE(B)	−120~−60	4	544~658	473	625	29	−93
5	MiuraSQV2Ah2 [38]	1TC(T)	−140~−80	4	542~709	461	602	23	−121
		2TC(T)	−100~−80	2	542~607	461	602	12	−121
		4TC(T)	−100~−80	2	542~607	461	602	12	−121
		0.4TC(T)	−140~−80	4	542~709	461	602	33	−121
		0.4TSE(B)	−140~−80	4	542~709	461	602	32	−121

*: Side-grooved specimens.

Table 4. K_{Jc} data used to construct the proposed tensile property-based MC: non-RPV steels.

Heat No.	Material	Specimen Type	Temps. (°C)	Num. of Temps.	σ_{YS} (MPa)	σ_{YSRT} (MPa)	σ_{BRT} (MPa)	Num. of Specimens	T_0 (°C)
6	GarciaS275JR [41]	1TC(T)	−50~−10	3	338~349	328	519	14	−26
7	GarciaS355J2 [41]	1TC(T)	−150~−100	3	426~528	375	558	13	−134
8	CiceroS460M [42]	0.6TSE(B)	−140~−100	3	597~686	473	595	14	−92
9	CiceroS690Q [42]	0.6TSE(B)	−140~−100	3	899~988	775	832	13	−111
10	MeshiiFY2017SCM440 [25,30]	0.9TSE(B)	−55~100	4	410~524	459	796	18	17
		0.5TSE(B)	−55~100	4	410~524	459	796	22	17
11	MeshiiFY2012S55C [6]	0.5TSE(B)	−25~20	3	394~444	394	707	17	27
12	MeshiiFY2016S55C [26,30]	0.9TSE(B)	−45~35	3	375~475	382	685	17	15
		0.5TSE(B)	−85~20	3	382~562	382	685	19	15

The objective variable was K_{Jc}. Assuming a direct relationship between the SED temperature dependence and that of K_{Jc}, σ_B temperature dependence was the first candidate explanatory parameter. However, considering that (i) σ_B/σ_{YS} temperature dependence is small, (ii) ferritic steel has a σ_{YS} temperature-dependent MC such as Z−A MC, and (iii) σ_B/σ_{YS} at RT is usually easily available, σ_B and σ_{YS} at RT, and σ_{YS} at K_{Jc} test temperatures and specimen width W were selected as the explanatory variables. To optimize the connection weight, 371 points, i.e., 70% of the 531 points in the known dataset, were used as the training data. The data were divided by "train_test_split" of Python's scikit-learn library. If the digits of the input value and output value to be learned are significantly different, the influence of variables with small digits may not be fully considered in learning. Therefore, in this study, the input values W, σ_{YS}, σ_{YSRT}, σ_{BRT}, and output value K_{Jc} were standardized, as shown in Equation (12).

$$\begin{pmatrix} W \text{ (mm)} \\ \sigma_{YS} \text{ (MPa)} \\ \sigma_{YSRT} \text{ (MPa)} \\ \sigma_{BRT} \text{ (MPa)} \\ K_{Jc} \left(\text{MPa}\cdot\text{m}^{1/2}\right) \end{pmatrix} \xrightarrow{\text{Normalized}} \begin{pmatrix} W/50 \\ \sigma_{YS}/550 \\ \sigma_{YSRT}/550 \\ \sigma_{BRT}/550 \\ K_{Jc}/100 \end{pmatrix} \quad (12)$$

Here, with reference to ASTM E1921, W was normalized using the width 50 mm of a 1T specimen, and the yield stress and tensile strength were normalized using the average value of 550 MPa of the yield stress of 275 to 825 MPa in the allowable temperature range targeted by the standard. K_{Jc} was normalized to a fracture toughness of value 100 MPa·m$^{1/2}$ at the reference temperature.

2.5. Fracture Toughness Prediction by the Constructed Learning Model

Table 5 presents a list of hyperparameters used for the machine learning model in this study. Using the data in Tables 2–4 and the parameters in Table 5, which is currently an invariant model, the coefficient of determination R^2 of the developed K_{Jc} predictor was 0.61 for the training data and 0.53 for the test data. Table 6 presents the explanation variables for predicting fracture toughness K_{Jc}.

Table 5. Hyperparameters used for the learning model.

Parameters	Value
Number of hidden layers	4
Number of hidden layer nodes	100, 50, 25, 10
Activation function	ReLU
Solver	Adam
α	0.01
η	0.001
β_1	0.9
β_2	0.999
ϵ	1.0×10^{-8}

Table 6. Explanatory variables for case studies applied to the developed tool.

Heat No.	Material	W (mm)	T (°C)	σ_{YSRT} (MPa)	σ_{YS} (MPa)	σ_{BRT} (MPa)
1	Miura SFVQ1A	20	−140, −120, −100, −80	454	695, 640, 607, 560	594
		50.8	−120, −100, −80, −60	454	640, 607, 560, 530	594
		101.6	−120, −100, −80, −60	454	640, 607, 560, 530	594
		203.2	−80, −100	454	607, 560	594
10	MeshiiFY2017SCM440	25	−55, 20, 60 100	459	524, 459, 435, 410	796
		46	−55, 20, 60, 100	459	524, 459, 435, 410	796

The input data (W, σ_{YS}, σ_{YSRT}, σ_{BRT}) for the developed K_{Jc} predictor and output window after its execution (the coefficient of determination R^2 and the predicted K_{Jc}) are shown in Figure 2. In Figure 3, the comparison of K_{Jc} of ASTM E1921 MC and predicted K_{Jc} by the predictor is shown. In Figure 3, the horizontal axis is T, the vertical axis $K_{Jc(1T)}$ is the test data, and the predicted K_{Jc} is converted to 1T thickness. The K_{Jc} of the ASTM E1921 MC is plotted as a black solid line, the K_{Jc} of the test data are plotted as open black symbols, and the predicted conditions listed in Table 6 are plotted as open red symbols. In Figure 3a, for RPV steel, both the K_{Jc} by the ASTM E1921 MC and the predicted K_{Jc} by this model are in agreement with the test results. However, in Figure 3b for SCM440, although the K_{Jc} by the ASTM E1921 MC significantly differs from the test results at high temperatures, the predicted K_{Jc} values by this model are in agreement with the test results.

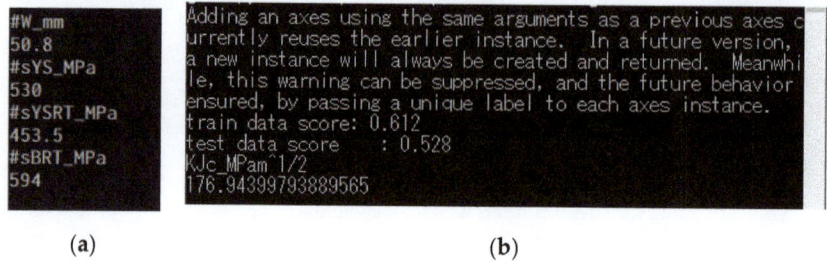

(a)   (b)

Figure 2. Input data (left figure) and window after execution (right figure). (**a**) Input data; (**b**) Output window.

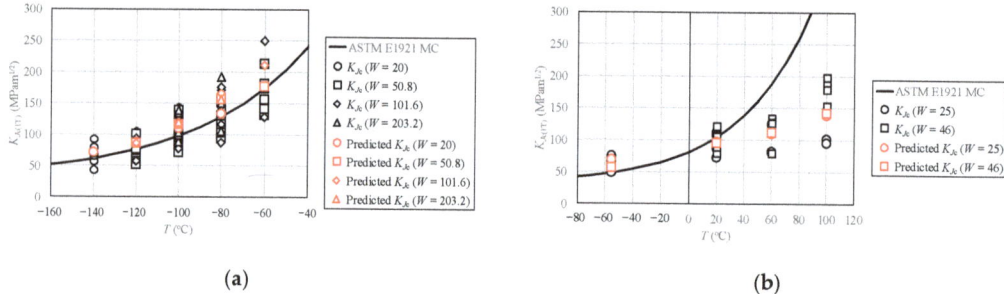

(a)   (b)

Figure 3. Comparison of K_{Jc} of ASTM E1921 MC (solid line) and predicted K_{Jc} by the predictor (open red symbols): Dataset used for training model and result of predicted K_{Jc}. (**a**) RPV steel (Miura SFVQ1A); (**b**) Meshii FY2017SCM440. K_{Jc} pre-dicted by the developed predictor accurately predicted K_{Jc} regardless of materials.

3. Discussion

By applying the ANN, a K_{Jc} predictor for ferritic steels that only requires tensile properties (i.e., σ_{YS} at the desired temperature for predicting K_{Jc}, and the RT values σ_{YSRT} and σ_{BRT}) were derived. This method eliminates the need for time- and material-consuming fracture toughness tests. The tool for predicting K_{Jc} by considering the specimen size and material properties is based on 531 fracture toughness test data values obtained from five RPV steel heats and seven non-RPV steel heats. The specimen sizes ranged from 0.4T to 4T to learn the size effect, the yield stress ranged from 328 to 775, and the tensile strength ranged from 519 to 832 to learn the material properties. The data range used in the training was equal to the application limit of the predictor. The developed K_{Jc} predictor successfully predicted training data with $R^2 = 0.61$ and test data with $R^2 = 0.53$.

To predict K_{Jc} at a specific temperature of interest, the user needs σ_{YS} at this temperature as well as σ_{YSRT} and σ_{BRT} at RT. If the material of interest is known to be well fitted by the Z–A σ_{YS} MC, the quantities for which test data are necessary for K_{Jc} prediction are only σ_{YSRT} and σ_{BRT}.

A considerable advantage of the proposed K_{Jc} predictor is that fracture toughness tests are not necessary to predict K_{Jc}. The key novel idea here is to use tensile properties (such as σ_{YS} and σ_B) and specimen size W.

Although the developed K_{Jc} predictor predicts one K_{Jc} for a combination of explanatory variables, the predicted K_{Jc} fracture probability is predicted by assuming the probability distribution of the data to be learned (e.g., Weibull distribution). It is also possible to evaluate it together, which is a future issue.

According to Tables 2–4, the $(\sigma_B/\sigma_{YS})_{RT}$ of non-RPV and RPV are different. Accepting Kirk's opinion that K_{Jc} and SED correspond, ASTM E1921 MC may deviate from non-RPV. However, this K_{Jc} predictor has an advantage in that it considers this. On this point, the developed K_{Jc} predictor, compared with the existing ASTM E1921 K_{Jc} MC, is expected to be especially effective in cases where σ_B/σ_{YS} of the material is larger than that of RPV steel.

The predictors that were generated and analyzed during the current study are available from the corresponding author upon reasonable request.

4. Conclusions

In this study, a tool was developed that can predict K_{Jc} for an arbitrary specimen size W and material properties (σ_{YSRT}, σ_{YS}, σ_{BRT}) via an ANN applied to 531 fracture toughness test data values. Currently, the conditions applicable to the tool are material properties ranging from σ_{YSRT} = 328 to 775 MPa, σ_{BRT} = 519 to 832 MPa, specimen size ranging from 0.4T to 4T and its types are CT and SEB. By using the tool developed through the application of data-driven ideas, it is possible to predict the fracture toughness at this temperature from the tensile test results and the specimen size at the target temperature of the fracture toughness without performing a fracture toughness test. In the future, it is planned to predict the predicted probability of fracture toughness.

Author Contributions: Conceptualization, T.M.; methodology, H.K. and Y.T.; software, H.K. and Y.T.; resource, T.M.; data curation, T.M.; writing-original draft preparation, K.I. and H.K. and Y.T. and T.M.; writing-review and editing, K.I. and T.M.; supervision, T.M.; project administration, T.M.; All authors have read and agreed to the published version of the manuscript.

Funding: This research received no external funding.

Data Availability Statement: The data that support the findings of this study are openly available in Appendix E at https://doi.org/10.1016/j.engfailanal.2020.104713 (accessed on 29 October 2021).

Acknowledgments: This work is part of the cooperative research between KOBELCO RESEARCH INSTITUTE, INC., and the University of Fukui. Support from both organizations is greatly appreciated.

Conflicts of Interest: The authors declare no conflict of interest.

Nomenclature

B	test specimen thickness
J	J-integral
K_{Jc}	fracture toughness
T	temperature (°C)
T_0	ASTM E1921 MC reference temperature (°C) for a 25 mm thick specimen with a fracture toughness of 100 MPa·m$^{1/2}$
W	specimen width
σ_{YS}, σ_B	yield (0.2% proof) and tensile strength
σ_{0ZA}	yield stress at the temperature T (°C) described by the Zerilli equation (i.e., Equation (11))
R^2	coefficient of determination
X_i	input value of MLP
a_j	activation unit of MLP
n	number of input value
k	number of activation unit
$f(X)$	output value of MLP
$w_{j,i}^h$	connection weight between input value X_i and activation unit a_j
ϕ	activation function
w_j^o	connection weight between activation unit a_j and output value $f(X)$
Y	teaching data
E	loss function
α	regularization strength of L^2 norm term
$w^{(t)}$	connection weight at timestep t in Adam
$m^{(t)}$	exponential moving averages of the gradient at timestep t in Adam
$v^{(t)}$	exponential moving averages of the squared gradient at timestep t in Adam
$\hat{m}^{(t)}$	bias-corrected first moment estimates at timestep t in Adam

$\hat{v}^{(t)}$	bias-corrected second raw moment estimates at timestep t in Adam
η	learning rate in Adam
ϵ	hyper parameter for numerical stability in Adam
β_1	hyper parameter for $m^{(t)}$ in Adam
β_2	hyper parameter for $v^{(t)}$ in Adam
μ_Y	average value of the true objective values

Abbreviations

ASTM	American Society for Testing and Materials
BCC	body-centered cubic
C(T)	compact tension; specimen type
DBT	ductile-to-brittle transition
MC	master curve
nT	notation used to indicate specimen thickness, where n is expressed in multiples of 25 mm
RPV	reactor pressure vessel
RT	room temperature
SE(B)	single-edge notched bend bar; specimen type
Z-A	Zerilli–Armstrong
SED	strain energy density
PCCV	pre-cracked Charpy V-notch; specimen type
MLP	multiplayer perceptron
ANN	artificial neural network
ReLU	rectified linear unit
Adam	adaptive moment estimation

References

1. James, P.M.; Ford, M.; Jivkov, A.P. A novel particle failure criterion for cleavage fracture modelling allowing measured brittle particle distributions. *Eng. Fract. Mech.* **2014**, *121–122*, 98–115. [CrossRef]
2. ASTM E1921-10. *Standard Test Method for Determination of Reference Temperature, T0, for Ferritic Steels in the Transition Range*; American Society for Testing and Materials: Philadelphia, PA, USA, 2010.
3. Odette, G.R.; He, M.Y. A cleavage toughness master curve model. *J. Nucl. Mater.* **2000**, *283*, 120–127. [CrossRef]
4. Ritchie, R.O.; Knott, J.F.; Rice, J.R. On the relationship between critical tensile stress and fracture toughness in mild steel. *J. Mech. Phys. Solids* **1973**, *21*, 395–410. [CrossRef]
5. Curry, D.A.; Knott, J.F. The relationship between fracture toughness and microstructure in the cleavage fracture of mild steel. *Met. Sci.* **1976**, *10*, 1–6. [CrossRef]
6. Ishihara, K.; Hamada, T.; Meshii, T. T-scaling method for stress distribution scaling under small-scale yielding and its application to the prediction of fracture toughness temperature dependence. *Theor. Appl. Fract. Mech.* **2017**, *90*, 182–192. [CrossRef]
7. Wallin, K. Irradiation damage effects on the fracture toughness transition curve shape for reactor pressure vessel steels. *Int. J. Press. Vessels Pip.* **1993**, *55*, 61–79. [CrossRef]
8. ASTM E1921-19b. *Standard Test Method for Determination of Reference Temperature, T0, for Ferritic Steels in the Transition Range*; American Society for Testing and Materials: Philadelphia, PA, USA, 2019.
9. Kirk, M.T.; Natishan, M.; Wagenhofer, M. Microstructural limits of applicability of the master curve. In *Fatigue and Fracture Mechanics*; Chona, R., Ed.; American Society for Testing and Materials: Philadelphia, PA, USA, 2001; Volume 32, pp. 3–16.
10. Kirk, M.T. *The Technical Basis for Application of the Master Curve to the Assessment of Nuclear Reactor Pressure Vessel Integrity*; ADAMS ML093540004; United States Nuclear Regulatory Commission: Washington, DC, USA, 2002.
11. Wallin, K. The size effect in KIC results. *Eng. Fract. Mech.* **1985**, *22*, 149–163. [CrossRef]
12. Dodds, R.H.; Anderson, T.L.; Kirk, M.T. A framework to correlate a/W ratio effects on elastic-plastic fracture toughness (J_c). *Int. J. Fract.* **1991**, *48*, 1–22. [CrossRef]
13. Kirk, M.T.; Dodds, R.H.; Anderson, T.L. An approximate technique for predicting size effects on cleavage fracture toughness (Jc) using the elastic T stress. In *Fracture Mechanics*; Landes, J.D., McCabe, D.E., Boulet, J.A.M., Eds.; American Society for Testing and Materials: Philadelphia, PA, USA, 1994; Volume 24, pp. 62–86.
14. Rathbun, H.J.; Odette, G.R.; He, M.Y.; Yamamoto, T. Influence of statistical and constraint loss size effects on cleavage fracture toughness in the transition—A model based analysis. *Eng. Fract. Mech.* **2006**, *73*, 2723–2747. [CrossRef]
15. Meshii, T.; Lu, K.; Takamura, R. A failure criterion to explain the test specimen thickness effect on fracture toughness in the transition temperature region. *Eng. Fract. Mech.* **2013**, *104*, 184–197. [CrossRef]
16. Meshii, T.; Yamaguchi, T. Applicability of the modified Ritchie-Knott-Rice failure criterion to transfer fracture toughness Jc of reactor pressure vessel steel using specimens of different thicknesses–possibility of deterministic approach to transfer the minimum Jc for specified specimen thicknesses. *Theor. Appl. Fract. Mech.* **2016**, *85*, 328–344. [CrossRef]

17. Anderson, T.L.; Stienstra, D.; Dodds, R.H. A theoretical framework for addressing fracture in the ductile-brittle transition region. In *Fracture Mechanics*; Landes, J.D., McCabe, D.E., Boulet, J.A.M., Eds.; American Society for Testing and Materials: Philadelphia, PA, USA, 1994; Volume 24, pp. 186–214.
18. Yang, S.; Chao, Y.J.; Sutton, M.A. Higher order asymptotic crack tip fields in a power-law hardening material. *Eng. Fract. Mech.* **1993**, *45*, 1–20. [CrossRef]
19. Meshii, T.; Tanaka, T. Experimental T_{33}-stress formulation of test specimen thickness effect on fracture toughness in the transition temperature region. *Eng. Fract. Mech.* **2010**, *77*, 867–877. [CrossRef]
20. Matvienko, Y.G. The effect of out-of-plane constraint in terms of the T-stress in connection with specimen thickness. *Theor. Appl. Fract. Mech.* **2015**, *80*, 49–56. [CrossRef]
21. Matvienko, Y.G. The effect of crack-tip constraint in some problems of fracture mechanics. *Eng. Fail. Anal.* **2020**, *110*, 104413. [CrossRef]
22. Beremin, F.M.; Pineau, A.; Mudry, F.; Devaux, J.C.; D'Escatha, Y.; Ledermann, P. A local criterion for cleavage fracture of a nuclear pressure vessel steel. *Metall. Mater. Trans. A* **1983**, *14*, 2277–2287. [CrossRef]
23. Khalili, A.; Kromp, K. Statistical properties of Weibull estimators. *J. Mater. Sci.* **1991**, *26*, 6741–6752. [CrossRef]
24. Zerilli, F.J.; Armstrong, R.W. Dislocation-mechanics-based constitutive relations for material dynamics calculations. *J. Appl. Phys.* **1987**, *61*, 1816–1825. [CrossRef]
25. Meshii, T. Failure of the ASTM E 1921 master curve to characterize the fracture toughness temperature dependence of ferritic steel and successful application of the stress distribution T-scaling method. *Theor. Appl. Fract. Mech.* **2019**, *100*, 354–361. [CrossRef]
26. Meshii, T. Spreadsheet-based method for predicting temperature dependence of fracture toughness in ductile-to-brittle temperature region. *Adv. Mech. Eng.* **2019**, *11*, 1–17. [CrossRef]
27. Zerbst, U.; Heerens, J.; Pfuff, M.; Wittkowsky, B.U.; Schwalbe, K.H. Engineering estimation of the lower bound toughness in the transition regime of ferritic steels. *Fatigue Fract. Eng. Mater. Struct.* **1998**, *21*, 1273–1278. [CrossRef]
28. Heerens, J.; Pfuff, M.; Hellmann, D.; Zerbst, U. The lower bound toughness procedure applied to the Euro fracture toughness dataset. *Eng. Fract. Mech.* **2002**, *69*, 483–495. [CrossRef]
29. Lu, K.; Meshii, T. Application of T_{33}-stress to predict the lower bound fracture toughness for increasing the test specimen thickness in the transition temperature region. *Adv. Mater. Sci. Eng.* **2014**, 1–8. [CrossRef]
30. Meshii, T. Characterization of fracture toughness based on yield stress and successful application to construct a lower-bound fracture toughness master curve. *Eng. Fail. Anal.* **2020**, *116*, 104713. [CrossRef]
31. Akbarzadeh, P.; Hadidi-Moud, S.; Goudarzi, A.M. Global equations for Weibull parameters in a ductile-to-brittle transition regime. *Nucl. Eng. Des.* **2009**, *239*, 1186–1192. [CrossRef]
32. Wenman, M.R. Fitting small data sets in the lower ductile-to-brittle transition region and lower shelf of ferritic steels. *Eng. Fract. Mech.* **2013**, *98*, 350–364. [CrossRef]
33. Meshii, T.; Yakushi, G.; Takagishi, Y.; Fujimoto, Y.; Ishihara, K. Quantitative comparison of the predictions of fracture toughness temperature dependence using ASTM E1921 master curve and stress distribution T-scaling methods. *Eng. Fail. Anal.* **2020**, *111*, 104458. [CrossRef]
34. Scikit-Learn. Available online: https://scikit-learn.org/stable/index.html (accessed on 13 October 2021).
35. Raschka, S.; Mrjalili, V. *Python Machine Learning*, 2nd ed.; Packt Publishing: Birmingham, UK, 2017.
36. Kingma, D.P.; Ba, J.L. ADAM: A METHOD FOR STOCHASTIC OPTIMIZATION. In Proceedings of the 3rd International Conference for Learning Representations, San Diego, CA, USA, 7–9 May 2015.
37. Rumelhart, D.E.; Hinton, G.E.; Williams, R.J. Learning representations by back-propagating errors. *Nature* **1986**, *323*, 533–536. [CrossRef]
38. Miura, N. Study on Fracture Toughness Evaluation Method for Reactor Pressure Vessel Steels by Master Curve Method Using Statistical Method. Ph.D. Thesis, The University of Tokyo, Tokyo, Japan, 2014. (In Japanese)
39. Gopalan, A.; Samal, M.K.; Chakravartty, J.K. Fracture toughness evaluation of 20MnMoNi55 pressure vessel steel in the ductile to brittle transition regime: Experiment & numerical simulations. *J. Nucl. Mater.* **2015**, *465*, 424–432. [CrossRef]
40. Rathbun, H.J.; Odette, G.R.; Yamamoto, T.; Lucas, G.E. Influence of statistical and constraint loss size effects on cleavage fracture toughness in the transition—A single variable experiment and database. *Eng. Fract. Mech.* **2006**, *73*, 134–158. [CrossRef]
41. García, T.; Cicero, S. Application of the master curve to ferritic steels in notched conditions. *Eng. Fail. Anal.* **2015**, *58*, 149–164. [CrossRef]
42. Cicero, S.; García, T.; Madrazo, V. Application and validation of the notch master curve in medium and high strength structural steels. *J. Mech. Sci. Technol.* **2015**, *29*, 4129–4142. [CrossRef]

Article

Research on the Corrosion Fatigue Property of 2524-T3 Aluminum Alloy

Chi Liu [1], Liyong Ma [2,3,*], Ziyong Zhang [3], Zhuo Fu [1] and Lijuan Liu [2]

[1] School of Mechanical and Electrical Engineering, Changsha University, Changsha 410022, China; liuchi@ccsu.edu.cn (C.L.); z20141061@ccsu.edu.cn (Z.F.)
[2] School of Mechanical Engineering, Hebei University of Architecture, Zhangjiakou 075031, China; zyq1292@hebiace.edu.cn
[3] School of Mechanical Engineering and Automation, Beihang University, Beijing 100091, China; ZY1907321@buaa.edu.cn
* Correspondence: maliyong@buaa.edu.cn; Tel.: +86-180-7516-1888

Abstract: The 2524-T3 aluminum alloy was subjected to fatigue tests under the conditions of $R = 0$, 3.5% NaCl corrosion solution, and the loading cycles of 10^6, and the S-N curve was obtained. The horizontal fatigue limit was 169 MPa, which is slightly higher than the longitudinal fatigue limit of 163 MPa. In addition, detailed microstructural analysis of the micro-morphological fatigue failure features was carried out. The influence mechanism of corrosion on the fatigue crack propagation of 2524-T3 aluminum alloy was discussed. The fatigue source characterized by cleavage and fracture mainly comes from corrosion pits, whose expansion direction is perpendicular to the principal stress direction. The stable propagation zone is characterized by strip fractures. The main feature of the fracture in the fracture zone is equiaxed dimples. The larger dimples are mixed with second-phase particles ranging in size from 1 to 5 μm. There is almost a one-to-one correspondence between the dimples and the second-phase particles. The fracture mechanism of 2524 alloy at this stage is transformed into a micro-holes connection mechanism, and the nucleation of micropores is mainly derived from the second-phase particles.

Keywords: 2524-T3 aluminum alloy; fatigue; corrosion; crack propagation; fracture

1. Introduction

2524-T3 aluminum alloy has the advantages of low density, high specific strength, excellent corrosion resistance, good formability, and low cost, so it is the main material in aircraft, vehicles, bridges, engineering equipment, and large pressure vessels [1–3]. In the service process, 2524-T3 aluminum alloy undergoes the alternating load for a long time, which has high requirements for the fatigue resistance of structural materials [4–6]. Especially in coastal areas and industrial areas with serious air pollution, structural parts are exposed to varying degrees of corrosive environments, such as salt spray and acid rain [4,7,8]. Under the interaction and synergy of alternating stress and corrosive environment, the fatigue resistance of components is significantly lower than that of ordinary mechanical fatigue, and the fatigue life is severely shortened. Corrosion fatigue is not a simple superposition of corrosion and fatigue damage but rather a process of synergy and promotion. Therefore, it has great destructive effects on the aluminum alloy structure [7,9].

Studies have been concentrated on the fatigue properties of aluminum alloys [10–12]. Zhang [13] conducted a fatigue test with A6005 aluminum alloy welded joints, demonstrating that the crack nucleation was a result of metallic oxides or discrete bar-like materials and that the crack propagation rates were inversely proportional to fractal dimension. C.S. Hattori et al. [14] studied the microstructure and fatigue properties of extruded aluminum alloys 7046 and 7108. The AA7046 displayed better tensile and fatigue properties than the AA7108. In addition, deep secondary cracks perpendicular to the growth direction of

the main crack were visible on all fracture surfaces. In the medium and high cycle fatigue tests of the AA7108 and AA7046, the cracks advanced in a perpendicular direction to the elongated grains resulting from the extrusion process.

In recent years, research studies have been conducted on the corrosion fatigue performance of aluminum alloy materials, mostly in the aerospace field. Ye et al. [15] performed the plasma electrolytic oxidation (PEO) on 7A85 aluminum alloy, and the influence on fatigue behavior in air and 3.5% NaCl solution was studied, demonstrating that PEO treatments significantly reduced the corrosion fatigue life of 7A85 aluminum alloy. R.K. et al. [16] studied the effects of corrosion on mechanical properties and fatigue life of 8011 aluminum alloy. Through tensile test and fatigue test, the research found that the corrosion had the severest destroy on the fatigue life of 8011 aluminum alloy structures. Zhang et al. [17] used ultrasonic nanocrystal surface modification (UNSM) to rejuvenate the fatigue performance of pre-corroded 7075-T651 aluminum alloy, finding that the fatigue life of the pre-corroded and UNSM treated specimens was twenty times higher than that of the only corroded specimens. Meanwhile, the reduction of the corroded surface layer and surface work hardening is beneficial for the fatigue performance rejuvenation of the pre-corroded alloy.

The relationship between the texture and grains with the fatigue properties of 2524 aluminum alloy were studied [1,3], demonstrating that the increasing of the intensity ratio of Cube to Brass texture is beneficial to the fatigue properties of 2524 aluminum alloy. Grain sizes among 50 and 100 mm exhibited high fatigue crack propagation resistances. When it comes to the effect of localized corrosion environment on fatigue properties of aluminum alloys, the studies reported recently [18–21] put their stress on the action of corrosion environment, especially the electrochemical effect on the fatigue life and properties.

In this paper, the corrosion fatigue properties of 2524-T3 aluminum alloy are investigated. The effect of 3.5% NaCl solution on the transverse and longitudinal corrosion fatigue properties of the alloy is studied and compared. The crack initiation, crack propagation, and the fracture processes of corrosion fatigue of 25,24-T3 aluminum alloy are analyzed. The effects of microstructure, such as the lengths of cracks and the widths of fatigue striations are discussed. By measuring the length of the crack and observing the width of the fatigue striations at different stages, the change in the crack growth rate is explained.

2. Materials and Experimental Methods

The experimental material is a 2524-T3 aluminum alloy sheet with a thickness of 4 mm, and its chemical composition is shown in Table 1. Rectangular specimens were selected. The longitudinal specimens were taken along the rolling direction, and the transverse specimens are taken perpendicular to the rolling direction. The width of the working section is 10 mm, the radius of the uniform transition chamfer is 120 mm, and the width of the clamping end is 30 mm. The engineering drawing of the sample for the corrosion fatigue experiment is shown in Figure 1.

Table 1. Chemical compositions of 2524-T3 aluminum alloy (%, mass fraction).

Mg	Zn	Cu	Cr	Ti	Mn	Si	Fe	Al
1.25	0.005	4.66	0.001	0.03	0.59	0.025	0.035	Bal.

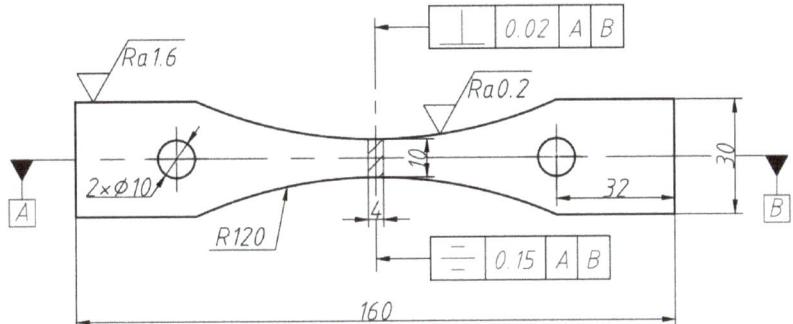

Figure 1. Engineering drawing of sample for corrosion fatigue experiment.

First, the samples were subjected to a universal tensile test on the MTS-LPS-204 universal testing machine (10 kN, MTS Industrial Co., Ltd., Eden Prairie, MN, USA) at a temperature of 25 °C. Second, the corrosion fatigue tests were conducted on the MTS-858 testing machine. The standard for the corrosion fatigue test is GB/T 20120.1-2006 (Corrosion of metals and alloys Corrosion fatigue test Part 1: Cyclic failure test) of National Standardization Administration of P.R. China. Before the test, the sample was fixed in the box mounted on the fatigue testing machine, as shown in Figure 2. Then, the box was filled with 3.5% NaCl solution, and the corrosion fatigue test was conducted after the box was sealed. The load was an axial sine wave, and the stress ratio $R = 0$ was selected. The number of cycles is set to 10^6, the test frequency is 3 Hz, and the stress levels are 5. The stress levels are selected according to the tensile test results, and the minimum number of samples is confirmed by the coefficient of variation.

Figure 2. Setup for corrosion fatigue test: (**a**) MTS-LPS-204 universal testing machine; (**b**) box to provide corrosion experiment.

3. Results and Discussion

3.1. S-N Curves

The universal tensile test shows that the tensile limit of 2524-T3 aluminum alloy is 475 MPa, and the first stress level of the fatigue test is selected as 190 MPa, which is 40% of the tensile limit. The stress levels of 2524-T3 aluminum alloy corrosion fatigue tests are 190 MPa, 180 MPa, 170 MPa, 160 MPa, and 150 MPa, respectively. The effective test results of horizontal and longitudinal corrosion fatigue test of materials under different stress levels are shown in Table 2. When the stress level was 150 MPa, the fracture did not appear after loading for 10^6 cycles, and the experiments finished.

Table 2. Data of horizontal and longitudinal corrosion fatigue test for 2524-T3 aluminum alloy.

Stress Level S_{max} (MPa)	Fatigue Lifetime of Horizontal Samples	Fatigue Lifetime of Longitudinal Samples
190	205,856	175,863
190	235,896	168,566
190	317,856	125,622
190	295,586	192,546
190	281,658	186,245
180	302,564	324,556
180	398,564	285,644
180	412,563	385,475
180	405,532	265,456
180	546,238	326,384
170	795,562	475,631
170	682,536	568,965
170	865,893	589,625
170	795,236	532,563
170	800,522	589,632
160	862,456	632,632
160	879,446	612,563
160	952,453	685,136
160	924,522	692,369
160	852,365	663,156
150	1,000,000	1,000,000
150	1,000,000	994,633
150	1,000,000	1,000,000
150	996,223	1,000,000
150	1,000,000	1,000,000

According to the test data in Table 2 and referring to Equations (1)–(3), the average value $\bar{x}$, standard deviation s, and coefficient of variation C of the sub-samples to judge the validity of the data are calculated. The least-squares method [22–24] is used to obtain a safety fatigue life curve with reliability and a confidence of 50%, as shown in Equations (1)–(3).

$$\bar{x} = \frac{1}{n}\sum_{i=1}^{n} \lg N_i = \hat{\mu} = \log \hat{N} \qquad (1)$$

$$\sigma = \sqrt{\frac{\sum_{i=1}^{n} x_i^2 - \frac{1}{n}\left(\sum_{i=1}^{n} x_i\right)^2}{n-1}} \qquad (2)$$

$$C = \frac{\delta_{max}\sqrt{n}}{p} \geq \frac{\sigma}{\bar{x}} \qquad (3)$$

where $\hat{\mu}$ is the population mean estimator; σ is the population standard deviation estimator; δ_{max} is the error limit, usually 5%; p is the probability density, which can be determined by looking up the table for n. The fitted curves are shown in Figure 3.

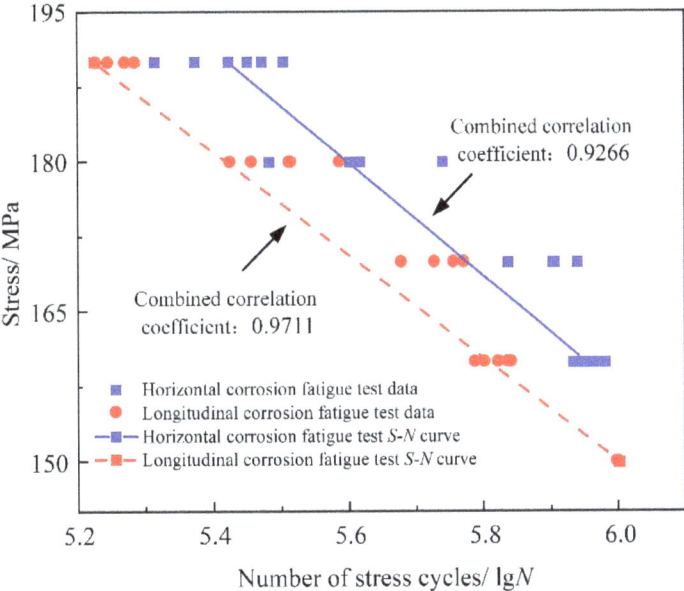

Figure 3. Horizontal and longitudinal corrosion fatigue tests for 2524-T3 aluminum alloy.

The S-N curve of the effective corrosion test under the confidence of 50% and the reliability of 50%, with different horizontal and longitudinal stress levels are shown in Equations (4) and (5), respectively. The combined correlation coefficient for the curves are 0.9266 and 0.9711, respectively.

$$S = 502.7457 - 55.2365 \lg N \quad (4)$$

$$S = 490.8512 - 49.5742 \lg N \quad (5)$$

In Equations (4) and (5), the horizontal fatigue limit of 2524-T3 aluminum alloy corresponding to 10^6 fatigue cycles is calculated to be 169 MPa, which is slightly higher than the longitudinal fatigue limit, which is 163 MPa. So, the fatigue lifetime of the longitudinal samples is slightly higher than that of the horizontal ones.

3.2. Fracture Analysis

Since the fracture process of the transverse and longitudinal specimens is similar, the longitudinal fracture of the specimen under 180 MPa is taken as an example to reveal the corrosion fatigue fracture process of the aluminum alloy.

Figure 4 is the macroscopic fatigue fracture morphology of the 2524-T3 aluminum alloy. It can be roughly divided into three areas from right to left:

- A—corrosion pitting and crack-oriented zone;
- B—fatigue crack propagation zone;
- C—rapid fatigue fracture zone.

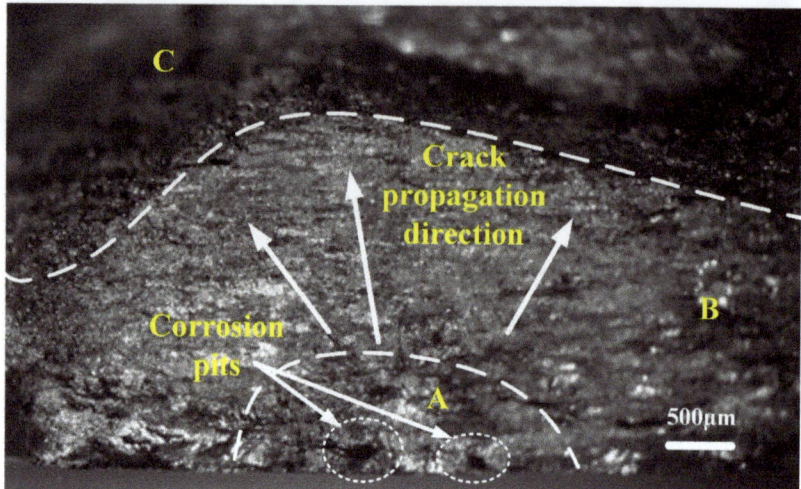

Figure 4. Morphologies of fracture specimen for 2524-T3 aluminum alloy at 180 MPa stress level.

As observed from the macroscopic fracture, the initial crack is directly emitted from the corrosion pitting of the sample. The crack propagation rate in Zone A is very slow, and the fatigue cracks are all nucleated from the surface. The corrosion pitting morphology can be observed in part of the shell lines in the fatigue source area, revealing that the fatigue cracks are initiated by the pitting pits caused by corrosion. From Zone A to depth, the crack enters the stable propagation stage, forming a flat Zone B. The fatigue crack propagation zone is bright, indicating a brittle intergranular fracture morphology, and numerous secondary cracks can be found in this area. As the crack continues to grow, the stress intensity factor at the crack tip increases, the fatigue crack propagates sharply, and the alloy enters the rapid fracture Zone C, which is rough and fibrous. The boundary between Zone B and Zone C is an obvious arc-shaped crack front. Near the crack front, the fracture has the mixture characteristics of crack propagation and plastic tearing.

In the crack propagation stage, the crack front is first generated near the center of the fracture. When the crack propagates to the vicinity of the surface of the specimen, the unbroken area cannot withstand the action of external force, and it breaks along the shear direction 45° to the cyclic stress. When the crack extends to Zone C, the plane strain fracture amount gradually decreases, and fracture occurs when it is stressed at an angle of 45°.

Figure 5 shows the 3D morphology of the fatigue fracture surface of the specimen in different crack regions. The dark red and dark blue in the figure indicate the highest and lowest positions, respectively.

Figure 5a shows the surface roughness of the fatigue source region. The fatigue section is relatively rough. At this time, the crack closure is mainly affected by the roughness. As the crack expands, it can be clearly seen that the peaks produced by the crack deflection are smoothed and flat, as shown in Figure 5b. It can be inferred that the mechanism that affects the crack closure in the medium and high stress zone has changed from roughness induction to plastic zone induction. The existence of the plastic zone at the crack tip triggers cracks in advance or mismatched contact. The original convex wave peaks are constantly squeezed during the cyclic loading process. Pressed and rubbed, it becomes flat and shiny. The plastic-induced crack closure leads to the contact between the crack surfaces in advance, and the crack-opening displacement is reduced, which indirectly reduces the driving force of crack growth, resulting in a decrease in the crack growth rate.

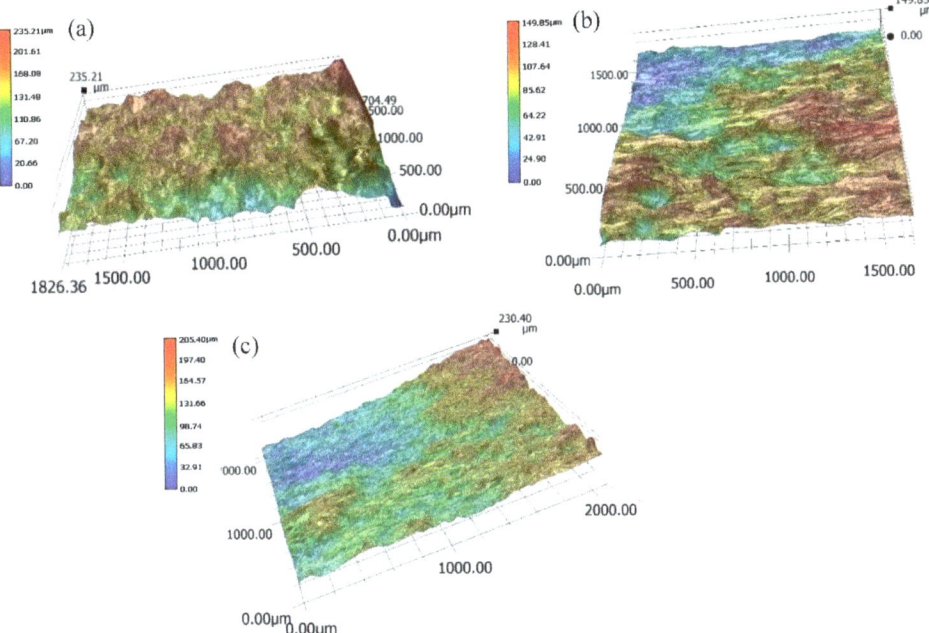

Figure 5. Three-dimensional (3D) morphology of the fatigue fracture surface of 2524-T3 aluminum alloy: (**a**) fatigue crack initiation zone; (**b**) crack propagation zone; (**c**) instantaneous failure zone.

In the later stage of crack propagation, the crack is in a rapid expansion state where the expansion and tearing are mixed. The crack may directly tear through several grains under each load, so the section in a small area appears very flat, as shown in Figure 5c. Meanwhile, at the upper right corner of the figure, the fracture is rising, and the feature of tearing appears.

3.2.1. Fatigue Source and the Initial Stage of Fatigue Crack Propagation

The SEM microscopic morphology of the fatigue source zone of the specimen after fracture at a stress level of 170 MPa is shown in Figure 6. The crack originates from the pits generated after the surface of the material is corroded, where stress concentration occurs under the action of fatigue loads. Excessive concentrated stress causes the dislocation movement of the material surface to intensify, forming a small slip zone where fatigue cracks are generated.

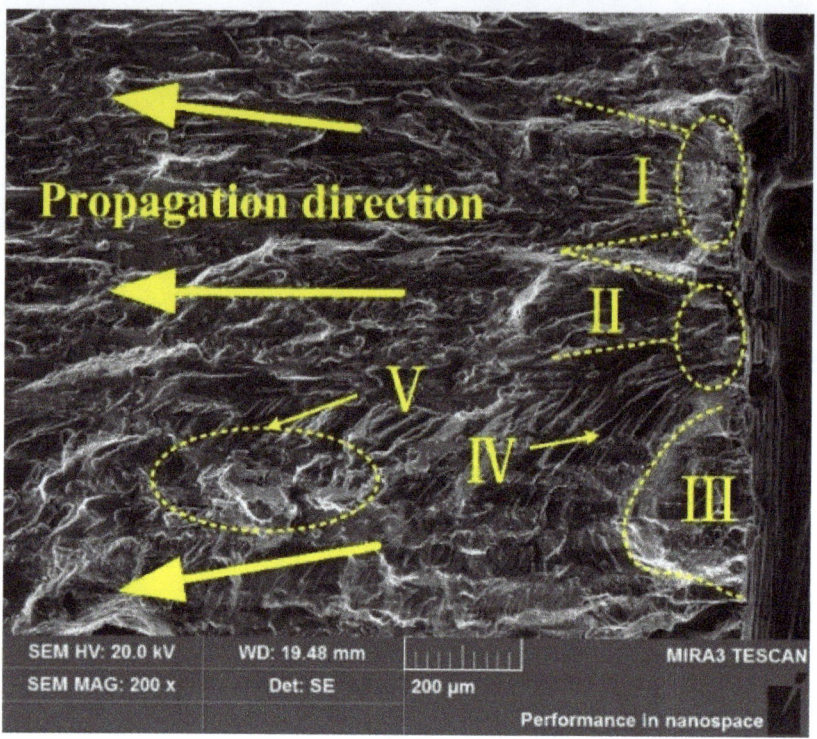

Figure 6. SEM micro-morphologies of fatigue source area.

In Figure 6, the fatigue source zone radiates toward the direction the cracks propagate and the crack front deviates in the propagation direction due to different resistances, so the crack starts to continue to expand along a series of planes with height differences, and the different fracture surfaces intersect to form steps, which constitute radial rays on the fatigue fracture. Near the corrosion pits, at Region I and Region II, the cracks originate at the corrosion pits, and there are obvious fan-shaped ray patterns along the crack propagation direction, showing ductile fracture characteristics. Region III may be affected by inclusions or twinning in the grains. The fracture characteristics feature a micro-zone cleavage mechanism under stress, and the crack source spreads around. After the main crack passes over both sides of the twinning at Region III, it continues to expand, forming a unique "tongue"-like pattern. The microscopic fracture morphology at Region IV presents the unique cleavage steps. Since the alloy has a high dislocation resistance at the grain boundary, when the crack propagates to the grain boundary, to minimize the energy consumed, the cleavage steps will expand along different crystal planes nearby. The face-centered cubic crystal structure (FCC) 2524-T3 aluminum alloy mainly slides along the direction <1$\bar{1}$0> on the sliding surface {111} [12]. As a result of the excellent toughness of 2524-T3 alloy, the proportion of the region with a brittle fracture is small. In addition, a stepped crack developed into the material at Region V, indicating that the crack tip has lateral slippage during the propagation of the main crack.

Figure 7 shows the fracture morphology of the 2524-T3 alloy at the initial stage of fatigue crack propagation. The fatigue crack propagation zone in the initial stage is flat, and the obvious quasi-cleavage fracture characteristics and wave-like pattern morphology can be seen, which is the main feature of damage in the corrosive environment. The crack propagates along a favorable direction relative to the maximum shear stress and extends to the depth of one or several grains. There are many relatively flat facets on the fractures with different heights. The small planes are connected by tear ridges (Figure 7a), demonstrating the crack propagations on different crystal planes. The tear ridges are deflected relative to the main crack direction, resulting in a certain angle difference in the main crack propagation direction. At the region shown by the cross-line in Figure 7b, the deflection angle is 32°. In addition, there are many micro-holes scattered on the cross-section, and the size and depth of the are different. It can be inferred that these micro-holes are caused by the coarse particles in the matrix being squeezed and stretched continuously during the crack propagation and then peeled off from the matrix.

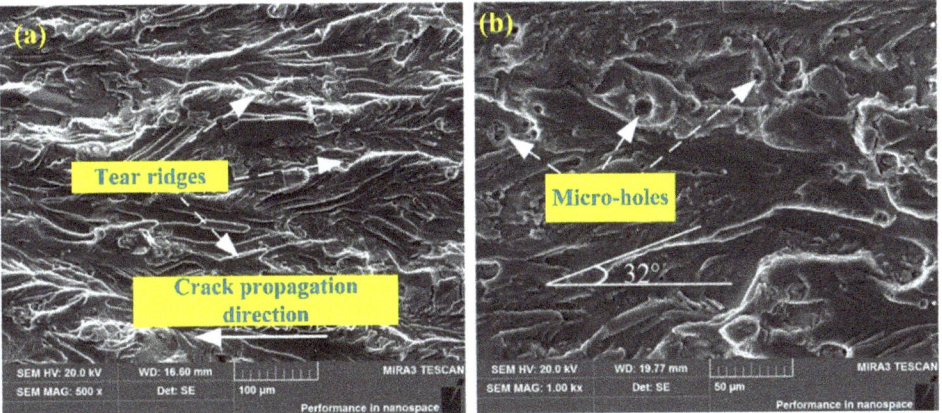

Figure 7. Morphology of fracture of 2524-T3 aluminum alloy in the initial stage of fatigue crack propagation: (**a**) tear ridges and crack propagation direction; (**b**) micro-holes and angle difference in the main crack propagation direction.

3.2.2. Stable Stage of Fatigue Crack Propagation

The fatigue life is mainly determined by the time of crack initiation and stable growth [25,26], and the crack propagation zone is the key feature of the fatigue fracture. Figure 6 shows the stable crack propagation area at a distance of 3 mm and 10 mm from the crack initiation end. Under ×1000 magnification, the main features in Figure 8a,c are similar. Along the crack propagation direction, the crack propagates in the form of transgranular fracture, and there are obvious crack edges and small planes on the fracture. This is because there are differences in the local orientation of cracks when they propagate inside the alloy. The resistance and growth rates encountered at the crack front are also different. The cracks continue to deviate as they propagate along their respective favorable slip surfaces, leaving different fracture surfaces intersecting. In Figure 8, the number of steps at the crack length of 3 mm is larger, indicating that the crack deflection is more frequent in the initial stage of crack propagation. The fracture contains a large number of unevenly distributed micro-holes. These micro-holes are formed by the gradual separation, breaking, and peeling of the unmelted coarse second-phase particles from the matrix under the action of cyclic stress.

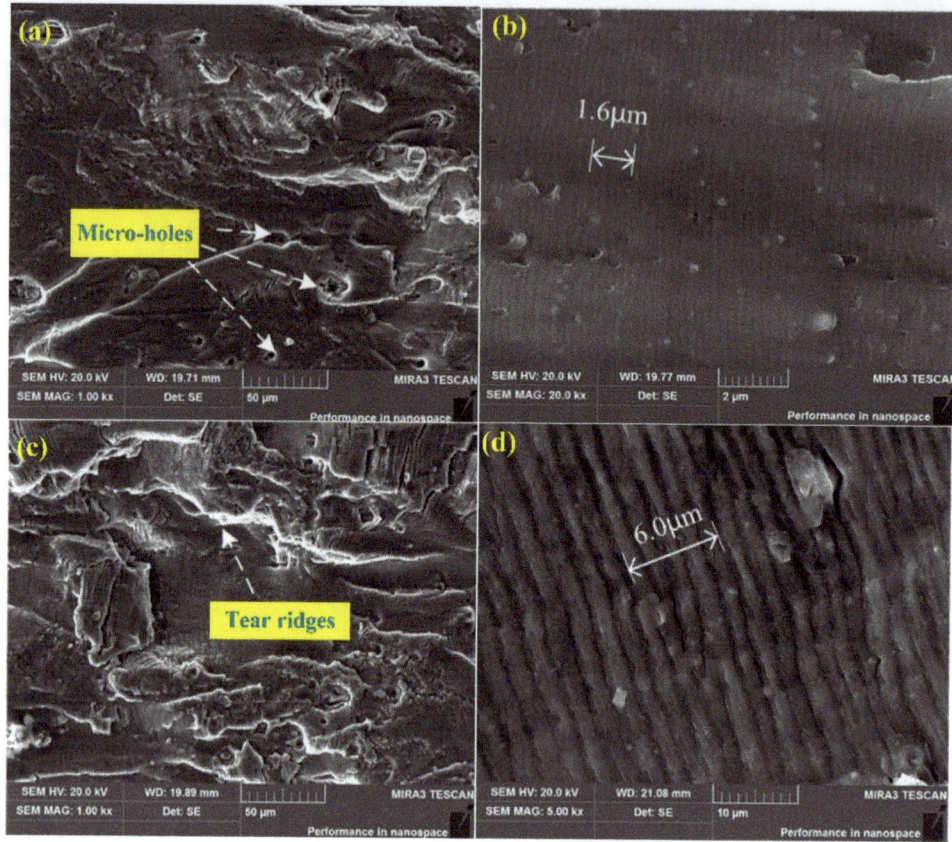

Figure 8. SEM images of stable crack propagation zone: (**a**,**b**) 3 mm from the crack initiation end; (**c**,**d**) 10 mm from the crack initiation end.

In Figure 8b,d, the relatively smooth areas of the fracture are further enlarged and observed, and obvious fatigue striations can be observed. Fatigue striations are the most typical microscopic feature of fatigue fracture. In Figure 8b,d, the thin and parallel fatigue striations are uniformly distributed, forming a group of parallel lines. The fatigue stripes on each small fracture plane are discontinuous and non-parallel, but their normal directions are along the crack propagation direction in the local fracture plane. In the process of crack propagation, the front of the crack is in an open plane strain state, and the crack propagation starts to proceed along the two slip systems at the same time or across. The double slip will cause the crack tip to be plastically passivated, i.e., fatigue striations. In Figure 8b, the width of the five fatigue striations is about 1.6 μm, and the distance of each ridge is about 0.32 μm, i.e., the microscopic crack propagation rate $da/dN = 0.32$ μm/cycle. The width of the five fatigue striations is significantly enlarged, about 6 μm, and the crack propagation rate is about three times higher than that at 3 mm, indicating that the spacing of the fatigue striations increases with the increase in the magnitude of the stress intensity factor.

Although the formation of fatigue striations is a localized process, the general propagation trend is along the propagation direction of macroscopic cracks. As shown in Figure 9, the fatigue striations on the fracture are not all distributed in the direction of crack propagation, and some will deviate from the direction of fatigue crack propagation. As shown in Figure 9a,b, when the fatigue crack crosses from one plane to another, it will also leave fatigue striations on planes with different directions and uneven heights [7]. 2524-T3 aluminum alloy is solid-solution strengthened and then subjected to natural aging treatment. The intragranular is mainly a shearable coherent phase GPB zone, and its cracks will propagate along the favorable slip surface. Due to the orientation differences between the two favorable slip planes in adjacent grains, the cracks continue to grow along the favorable slip plane after growing across the grain boundary, and they gradually deflect. The fatigue striations on both sides of the grain boundary present an angle, but there is no clear change in the size of the fatigue striations on both sides. The deflection process of fatigue cracks consumes the deformation and storage energy under the action of cyclic stress, effectively reduces stress concentration, and improves the fatigue resistance of the aluminum alloy. As shown in Figure 9c, when the fatigue crack encounters the twinning boundary, it is swallowed by the twinning boundary and continues to expand [27,28]. The fatigue striations show a symmetrical relationship along the twin boundary, but the widths are unchanged, indicating that the co-lattice twinning interface energy in 2524 aluminum alloy is relatively low, and the effect on the crack propagation rate is not obvious. When the orientation difference of adjacent grains is not large, the crack can expand through the grain boundary and enter another grain without changing the expansion angle too much, which is conducive to the transgranular expansion of the crack and the formation of obvious straight cracks. When the crack propagates in the grain, it will preferentially propagate along a certain cleavage plane, and the propagation path may form a "Z"-shaped crack, as shown in Figure 9d.

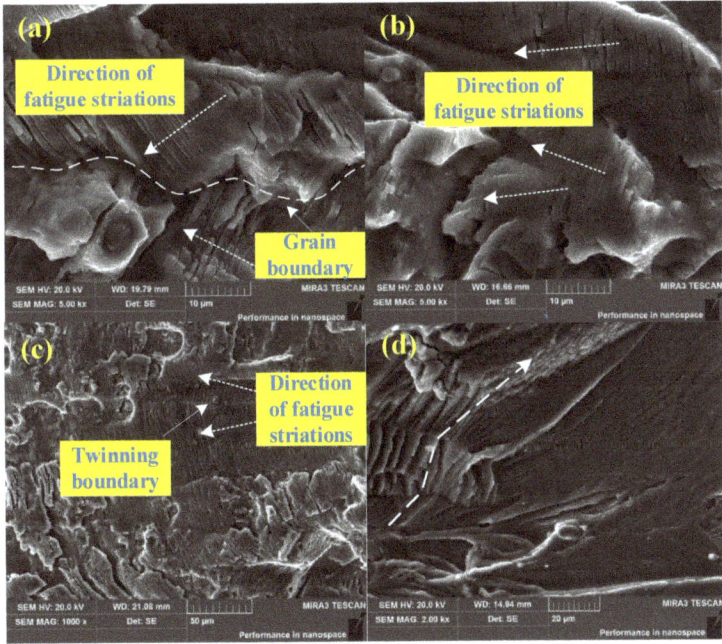

Figure 9. SEM images of different orientations in the stable crack propagation zone: (**a**,**b**) fatigue fringes on adjacent grains with different orientations; (**c**) twinning and fatigue fringes on both sides; (**d**) "Z"-shaped crack.

The 2524-T3 alloy sheet contains a large number of micron-sized second-phase particles which are generally Al_2CuMg phase and Fe-rich phase. The impurity phase will leave obvious features on the fatigue fracture. As shown in Region A in Figure 10a, there are a large number of broken coarse second-phase particles and holes on the crack propagation path. This is because under the action of cyclic stress, some of the unmelted, coarse second-phase particles are torn during the crack propagation process to form broken particles, and some are separated from the matrix and leave holes. These broken particles and holes provide a preferential path for crack propagation. In Region B, because the coarse second-phase particles break under the action of cyclic stress, the fatigue striations choose to bypass the particles for expansion, indicating that the cracks tend to expand in the direction of more inclusions [29,30], bridging larger debonded inclusions, and thus, the fatigue resistance of the material is weakened.

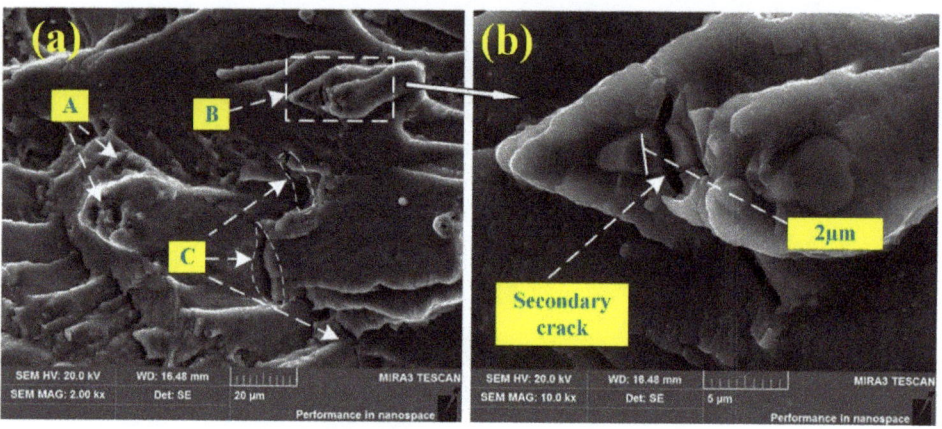

Figure 10. SEM images of coarse particles and secondary cracks in the stable crack propagation zone: (**a**) coarse particles; (**b**) secondary cracks.

In addition, another important feature of 2524 aluminum alloy during the crack propagation process—secondary cracking—was observed on the fracture. As shown in Region C in Figure 10a, there are a large number of secondary cracks distributed along the direction of the glare on the fatigue fracture. They are cracks that expand from the surface of the fracture to the inside, which is distributed intermittently on the fracture, and the directions are often parallel to the fatigue striations, but the depths are much greater than the depths of the striations. Some secondary cracks are initiated and propagated along the second-phase particles (in Figure 10b). Golden et al. [31] explained this phenomenon, demonstrating that stress concentration leads to the accumulation of dislocations that generate along the weak position of the phase interface, and the stress is relieved after the second cracking. From the point of view of the energy method, the fatigue damage process is the accumulation of strain energy in the plastic zone under cyclic stress [32,33]. The formation of secondary cracks is beneficial to release the energy at the crack tip and slow down the crack propagation rate to a certain extent.

3.2.3. Stage of Rapid Fatigue Crack Propagation and Fatigue Fracture

When the stress intensity factor amplitude of the crack tip is $\Delta K \approx 25$ MPa·m$^{1/2}$, the crack enters the rapid growth zone, and the crack tip expands to the position shown by the dotted line of the crack front in Figure 3. Near the boundary, there is a mixed fracture morphology of the transition between the crack propagation zone and the transient zone. This area was observed under a high-power SEM electron microscope, and the results are shown in Figure 11.

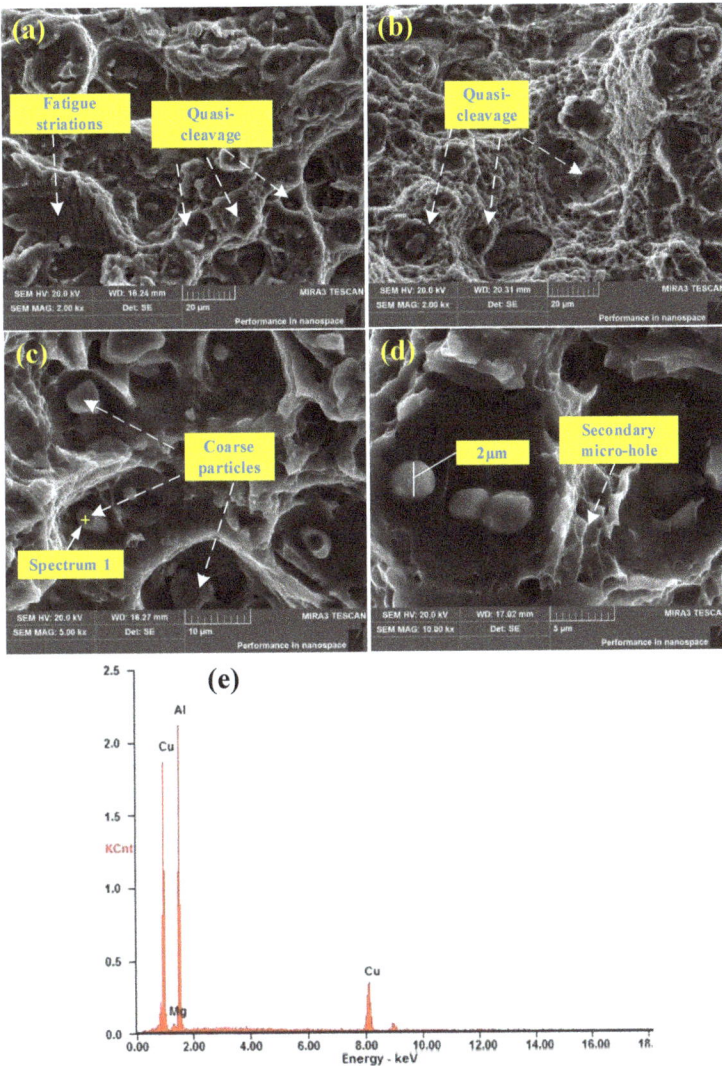

Figure 11. Microscopic morphology of the fatigue rapid growth zone: (**a**) crack propagation and fracture transition zone; (**b**) instantaneous failure zone; (**c**,**d**) enlarged views of dimples, (**e**) EDX map of the coarse particles.

On the right side of the dividing line (shown in Figure 11a), the crack has just transferred from the stable propagation zone to the rapid propagation zone. There are a few fatigue striations on the fracture with a large distance of 2 μm. The fracture morphology shows the characteristics of a mixture of fatigue bands and dimples. On the left side of the dividing line, the material begins to lose stability, and fracture occurs. The microscopic morphology begins to change from a mixed dimple-fatigue striation zone to a dimple tear zone (shown in Figure 11b). Microscopically, the fracture of the dimple is honeycomb-like, composed of many holes and small pits. The fracture morphology is characterized by quasi-static ductile tensile fracture [34], and the fracture mechanism is a microporous connection.

Dimples are the main feature of fatigue fracture of 2524 alloy at this stage. Due to the normal tensile stress of the specimen in the test, the microscopic voids in the alloy will grow at the same rate along the three sides of the space, thus forming the equiaxed dimples in the figure. The size of the dimples is not the same. The large dimples are full of small dimples. A large number of sliding steps and tearing edges can be seen at the boundaries of the dimples. This indicates that during the rapid expansion stage, the 2524 alloy fractured after a relatively large plastic deformation. These characteristics were observed in a previous study by Magnusen et al. [10]. The process of dimple fracture is divided into three stages: microporous nucleation–growth–polymerization. In Figure 11b, it can be observed that the larger-sized dimples are all mixed with second-phase particles, with sizes ranging from 1 to 5 μm. The second-phase particles have an almost one-to-one relationship, which verifies the view that the second-phase particles are the source of micropore nucleation.

In the 2524 alloy, there are mainly impurity phases such as Mg_2Si, Al_7Cu_2Fe, $Al_{12}FeSi$, $(MgFe)_3SiAl_{12}$, $(FeMn)Al_3$, $(FeMn)Al_6$, and S (Al_2CuMg) equilibrium phase. These coarse impurity phases are very brittle, and they are easily separated from the matrix under stress or crack to form micro-holes (Figure 11c). Some strengthening phases (Al_2CuMg) that are relatively firmly combined with the matrix will eventually produce micro-holes due to inconsistent plastic deformation with the matrix under the severe stress concentration in the later stage of crack propagation, as shown in Figure 11d. Excessive coarse second phases inside the alloy will severely affect the fracture toughness of the alloy. Therefore, to improve the fracture toughness of the alloy, the size and quantity of the coarse second phase should be significantly reduced. Figure 11e is an EDX map of Spectrum 1 of the coarse particle in Figure 11c. The chemical composition of Spectrum 1 is 46.9% Al, 52.5% Cu, 0.6% Mg, which can be judged as Al_2Cu phase, which is brittle. Under the action of alternating loads, the coarse second phase cannot simultaneously deform in coordination with the matrix, and dislocations will continue to accumulate at the particle–phase interface and cause stress concentration. When the peak stress in the local area exceeds the fracture strength of the alloy, the coarse phase will be separated from the matrix at the interface and peeled from the aluminum matrix to form cavities, which is also claimed in Ref. [35].

4. Conclusions

- Under the condition of $R = 0$, 3.5% NaCl corrosion solution, and the loading cycles of 10^6, the horizontal and longitudinal corrosion fatigue limits of the 2524-T3 aluminum alloy are 495 and 523 MPa, respectively. The horizontal fatigue corrosion performance is slightly better than the performance of longitudinal corrosion.
- In the morphology of the fatigue fracture, the crack closure of the fatigue source region is mainly affected by the roughness. As the crack expands, the mechanism that affects the crack closure in the medium and high-stress zone has changed from roughness induction to plastic zone induction. In the later stage of crack propagation, the crack may directly tear through several grains under each load, so the section in a small area appears very flat.
- Fatigue cracks mainly originate from corrosion pits. In the initial stage of crack propagation, the fracture surface shows a mixed characteristic of ductile fracture and cleavage fracture. The cracks propagate on different crystal planes, forming small steps with different heights and deflected tear ridges.
- During the stable propagation stage of fatigue cracks, the cracks mainly propagate through the double-slip mechanism. At this stage, clear, smooth, and parallel plastic fatigue striations can be observed. The width of fatigue striations increases with the crack length or the amplitude of the stress intensity factor, and the striation direction is perpendicular to the local crack propagation direction. When the fatigue striations pass through the coarse second-phase particles, they expand by bypassing the particles, indicating that the internal cracks of the alloy tend to expand in the direction of more inclusions, bridging larger debonded inclusions, and thus, the alloy's fatigue resistance is weakened. In the rapid propagation/fracture stage, the

crack propagation/fracture mechanism is transformed into a micro-holes connection mechanism. The main characteristic morphology at this stage is equiaxed dimples. Larger dimples are mixed with second-phase particles ranging in size from 1 to 5 μm. The relationship between dimples and second-phase particles is almost one-to-one, indicating that the nucleation of micropores mainly comes from the second-phase particles in the alloy.

Author Contributions: Conceptualization, C.L. and L.M.; methodology, C.L. and L.M.; validation, C.L., L.M. and Z.Z.; formal analysis, L.L. and Z.F.; investigation, C.L., L.M. and L.L.; resources, C.L.; data curation, C.L.; writing—original draft preparation, C.L.; writing—review and editing, L.M.; project administration, C.L., L.M., Z.F. and L.L. All authors have read and agreed to the published version of the manuscript.

Funding: This research was funded by the Scientific Research Fund of Hunan Provincial Education Department (grant number 20C0168 and 20B068), Changsha Municipal Natural Science Foundation (kq2007085), the Basic Scientific Research Business Project of Hebei University of Architecture (2021QNJS08), the Science and Technology Research and Development Command Plan of Zhangjiakou (1911031A), and "14th Five-Year Plan" Project of Hebei Higher Education Association (GJXH2021-109).

Data Availability Statement: The data presented in this study are available on request from the corresponding author.

Acknowledgments: The authors acknowledge Xiaohong Sun for providing support in experiments.

Conflicts of Interest: The authors declare no conflict of interest.

References

1. Shen, F.; Yi, D.; Wang, B.; Liu, H.; Jiang, Y.; Tang, C.; Jiang, B. Semi-quantitative evaluation of texture components and anisotropy of the yield strength in 2524 T3 alloy sheets. *Mater. Sci. Eng. A* **2016**, *675*, 386–395. [CrossRef]
2. Xiong, J.; Peng, X.; Shi, J.; Wang, Y.; Sun, J.; Liu, X.; Li, J. Numerical simulation of thermal cycle and void closing during friction stir spot welding of AA-2524 at different rotational speeds. *Mater. Charact.* **2021**, *174*, 110984. [CrossRef]
3. Yin, D.; Liu, H.; Chen, Y.; Yi, D.; Wang, B.; Wang, B.; Shen, F.; Fu, S.; Tang, C.; Pan, S. Effect of grain size on fatigue-crack propagation in 2524 aluminium alloy. *Int. J. Fatigue* **2016**, *84*, 9–16. [CrossRef]
4. Costenaro, H.; Lanzutti, A.; Paint, Y.; Fedrizzi, L.; Terada, M.; de Melo, H.G.; Olivier, M.-G. Corrosion resistance of 2524 Al alloy anodized in tartaric-sulphuric acid at different voltages and protected with a TEOS-GPTMS hybrid sol-gel coating. *Surf. Coat. Technol.* **2017**, *324*, 438–450. [CrossRef]
5. Baptista, C.A.R.P.; Adib, A.M.L.; Torres, M.A.S.; Pastoukhov, V.A. Describing fatigue crack propagation and load ratio effects in Al 2524 T3 alloy with an enhanced exponential model. *Mech. Mater.* **2012**, *51*, 66–73. [CrossRef]
6. Xu, Y.; Zhan, L.; Li, W. Effect of pre-strain on creep aging behavior of 2524 aluminum alloy. *J. Alloy. Compd.* **2017**, *691*, 564–571. [CrossRef]
7. Moreto, J.; Gelamo, R.; Nascimento, J.; Taryba, M.; Fernandes, J. Improving the corrosion protection of 2524-T3-Al alloy through reactive sputtering Nb2O5 coatings. *Appl. Surf. Sci.* **2021**, *556*, 149750. [CrossRef]
8. Srivatsan, T.S.; Kolar, D.; Magnusen, P. Influence of temperature on cyclic stress response, strain resistance, and fracture behavior of aluminum alloy 2524. *Mater. Sci. Eng. A* **2001**, *314*, 118–130. [CrossRef]
9. Sonsino, C.M. Consideration of Salt-Corrosion Fatigue for Lightweight Design and Proof of Aluminium Safety Components in Vehicle Applications. *Int. J. Fatigue* **2021**, *154*, 106406. [CrossRef]
10. Magnusen, T. The cyclic fatigue and final fracture behavior of aluminum alloy 2524. *Mater. Des.* **2002**, *23*, 129–139.
11. Shen, F.; Yi, D.; Jiang, Y.; Wang, B.; Liu, H.; Tang, C.; Shou, W. Semi-quantitative evaluation of texture components and fatigue properties in 2524 T3 aluminum alloy sheets. *Mater. Sci. Eng. A* **2016**, *657*, 15–25. [CrossRef]
12. Maduro, L.P.; Baptista, C.A.R.P.; Torres, M.A.S.; Souza, R.C. Modeling the growth of LT and TL-oriented fatigue cracks in longitudinally and transversely pre-strained Al 2524-T3 alloy. *Procedia Eng.* **2011**, *10*, 1214–1219. [CrossRef]
13. Zhang, Z. Quantitative characterization on fatigue fracture features of A6005 aluminum alloy welded joints. *Eng. Fail. Anal.* **2021**, *129*, 105687. [CrossRef]
14. Hattori, C.S.; Almeida, G.F.C.; Gonçalves, R.L.P.; Santos, R.G.; Souza, R.C.; da Silva, W.C., Jr.; Cunali, J.R.C., Jr.; Couto, A.A. Microstructure and Fatigue Properties of Extruded Aluminum Alloys 7046 and 7108 for Automotive Applications. *J. Mater. Res. Technol.* **2021**, *14*, 2970–2981. [CrossRef]
15. Ye, Z.; Liu, D.; Zhang, X.; Wu, Z.; Long, F. Influence of combined shot peening and PEO treatment on corrosion fatigue behavior of 7A85 aluminum alloy. *Appl. Surf. Sci.* **2019**, *486*, 72–79. [CrossRef]

16. Mishra, R.K. Study the effect of pre-corrosion on mechanical properties and fatigue life of aluminum alloy 8011. *Mater. Today Proc.* **2020**, *25*, 602–609. [CrossRef]
17. Zhang, R.; Zhao, W.; Zhang, H.; Yang, W.; Wang, G.-X.; Dong, Y.; Ye, C. Fatigue Performance Rejuvenation of Corroded 7075-T651 Aluminum Alloy through Ultrasonic Nanocrystal Surface Modification. *Int. J. Fatigue* **2021**, *153*, 106463. [CrossRef]
18. Moreto, J.A.; dos Santos, M.S.; Ferreira, M.O.A.; Carvalho, G.S.; Gelamo, R.V.; Aoki, I.V.; Taryba, M.; Filho, W.W.B.; Fernandes, J.C.S. Corrosion and corrosion-fatigue synergism on the base metal and nugget zone of the 2524-T3 Al alloy joined by FSW process. *Corros. Sci.* **2021**, *182*, 109253. [CrossRef]
19. Moreto, J.A.; Broday, E.E.; Rossino, L.S.; Fernandes, J.C.S.; Bose Filho, W.W. Effect of Localized Corrosion on Fatigue–Crack Growth in 2524-T3 and 2198-T851 Aluminum Alloys Used as Aircraft Materials. *J. Mater. Eng. Perform.* **2018**, *27*, 1917–1926. [CrossRef]
20. Rangel, U.; Borges, R.; Oliveira, D.A.; de Almeida, L.S.; Gelamo, R.V.; Siqueira, J.R., Jr.; Rossino, L.S.; Moreto, J.A. Corrosion and Micro-abrasive Wear Behaviour of 2524-T3 Aluminium Alloy with PAni-NPs/PSS LbL Coating. *Mater. Res.* **2019**, *22*. [CrossRef]
21. Chlistovsky, R.M.; Heffernan, P.J.; Duquesnay, D.L. Corrosion-fatigue behaviour of 7075-T651 aluminum alloy subjected to periodic overloads. *Int. J. Fatigue* **2007**, *29*, 1941–1949. [CrossRef]
22. Jiang, F.; Ding, Y.; Song, Y.; Geng, F.; Wang, Z. Digital Twin-driven framework for fatigue life prediction of steel bridges using a probabilistic multiscale model: Application to segmental orthotropic steel deck specimen. *Eng. Struct.* **2021**, *241*, 112461. [CrossRef]
23. Omrani, A.; Langlois, S.; Van Dyke, P.; Lalonde, S.; Karganroudi, S.S.; Dieng, L. Fretting fatigue life assessment of overhead conductors using a clamp/conductor numerical model and biaxial fretting fatigue tests on individual wires. *Fatigue Fract. Eng. Mater. Struct.* **2021**, *44*, 1498–1514. [CrossRef]
24. Liu, X.R.; Sun, Q. An improved macro—Micro-two-scale model to predict high-cycle fatigue life under variable amplitude loading. *Contin. Mech. Thermodyn.* **2021**, *33*, 803–816. [CrossRef]
25. Chen, Y.; Zheng, Z.; Cai, B.; Xu, J. Initiation and Propagation Behavior of Fatigue Cracks in 2197(Al-Li)-T851 Alloy. *Rare Met. Mater. Eng.* **2011**, *40*, 1926–1930.
26. Tanaka, K. The Propagation of Small Fatigue Cracks. *J. Soc. Mater. Sci. Jpn.* **1989**, 869–887. [CrossRef]
27. Schur, E.; Afari, N.; Goldberg, J.; Buchwald, D.; Sullivan, P.F. Twin analyses of fatigue. *Twin Res. Hum. Genet.* **2007**, *10*, 729–733. [CrossRef]
28. Stein, C.A. The Role of Twin Boundaries in Fatigue Microcrack Initiation in an Advanced Nickel-Based Alloy. Ph.D. Thesis, Carnegie Mellon University, Pittsburgh, PA, USA, 2015.
29. Chen, Y.; Pan, S.; Zhou, M.; Yi, D.; Xu, D.; Xu, Y. Effects of inclusions, grain boundaries and grain orientations on the fatigue crack initiation and propagation behavior of 2524-T3 Al alloy. *Mater. Sci. Eng. A* **2013**, *580*, 150–158. [CrossRef]
30. Pippan, R.; Zelger, C.; Gach, E.; Bichler, C.; Weinhandl, H. On the mechanism of fatigue crack propagation in ductile metallic materials. *Fatigue Fract. Eng. Mater. Struct.* **2015**, *34*, 1–16. [CrossRef]
31. Golden, P.J.; Grandt, A.; Bray, G. A comparison of fatigue crack formation at holes in 2024-T3 and 2524-T3 aluminum alloy specimens. *Int. J. Fatigue* **1999**, *21*, S211–S219. [CrossRef]
32. Benam, A.S.; Yazdani, S.; Avishan, B. Effect of shot peening process on fatigue behavior of an alloyed austempered ductile iron. *China Foundry* **2011**, *8*, 325–330.
33. Azadi, M.; Shirazabad, M.M. Heat treatment effect on thermo-mechanical fatigue and low cycle fatigue behaviors of A356.0. *Alum. Alloy. Mater. Des.* **2013**, *45*, 279–285. [CrossRef]
34. Liu, Z.; Wang, J.; Gao, H.; Gao, L. Biaxial fatigue crack propagation behavior of ultrahigh molecular weight polyethylene reinforced by carbon nanofibers and hydroxyapatite. *J. Biomed. Mater. Res. Part B Appl. Biomater.* **2020**, *108*, 1603–1615. [CrossRef]
35. Chen, Y.; Tang, Z.; Pan, S.; Liu, W.; Song, Y.; Liu, Y.; Zhu, B. The fatigue crack growth behaviour of 2524–T3 aluminium alloy in an Al_2O_3 particle environment. *Fatigue Fract. Eng. Mater. Struct.* **2020**, *43*, 2376–2389. [CrossRef]

Editorial

Fracture Mechanics and Fatigue Design in Metallic Materials

Dariusz Rozumek

Department of Mechanics and Machine Design, Opole University of Technology, Mikolajczyka 5, 45-271 Opole, Poland; d.rozumek@po.edu.pl; Tel.: +48-77-449-8410

1. Introduction and Scope

Devices, working structures and their elements are subjected to the influence of various loads. These can be static, cyclic or dynamic loads. The accumulation of damage and the development of fatigue cracks under the influence of loads is a common phenomenon that occurs in metals. To slow down crack growth and ensure an adequate level of safety and the optimal durability of structural elements, experimental tests and simulations are required to determine the influence of various factors. Such factors include, among others, the impact of microstructures, voids, notches, the environment, etc. Research carried out in this field and the results obtained are necessary to guide development toward the receipt of new and advanced materials that meet the requirements of the designers. This Special Issue aims to provide the data, models and tools necessary to provide structural integrity and perform lifetime prediction based on the stress (strain) state and, finally, the increase in fatigue cracks in the material, which would result in the application of advanced mathematical, numerical and experimental techniques.

2. Contributions

Fracture mechanics are present in most structures that work cyclically, e.g., in the automotive or aviation industry. To extend the life of structures, they must be properly fatigue-proofed and made of appropriate materials. This Special Issue shows the fatigue behavior of various alloys and the conditions under which these alloys work. A paper by Sharma et al. [1] reviews the research and development in the field of fatigue damage, focusing on the very high cycle fatigue (VHCF) of metals, alloys and steels. In addition, they showed the influence of various defects, crack initiation sites, fatigue models and simulation studies to understand the crack development in VHCF regimes. A paper by Wang et al. [2] investigates the influence of the crack behavior propagation process in welded joints and sheds light on the mechanism of their branching, and a paper by Wei Xu et al. [3] proposes an ultra-high-frequency (UHF) fatigue test of a titanium alloy TA11 based on an electrodynamic shaker to develop a feasible testing method in the VHCF regime. The results from UHF tests data show good consistency with those from the axial-loading fatigue and rotating bending fatigue tests. Moreover, the fatigue life obtained from an ultrasonic fatigue test with the loading frequency of 20 kHz is significantly higher than all the other fatigue test results.

Artola et al. [4] investigated the impact of quench and tempering and hot-dip galvanizing on the hydrogen embrittlement behavior of a high-strength steel. Slow-strain-rate tensile testing was employed to assess this influence. Two sets of specimens were tested, both in-air and immersed in synthetic seawater. It was found that the risk of rupture only arises due to hydrogen re-embrittlement in wet service.

The closure of the crack was discussed in three articles [5–7]. Zakavi et al. [5] presents new tools to evaluate the crack front shape of through-the-thickness cracks propagating in plates under quasi-steady-state conditions. A numerical approach incorporating simplified phenomenological models of plasticity-induced crack closure was developed and validated against experimental results.

Citation: Rozumek, D. Fracture Mechanics and Fatigue Design in Metallic Materials. *Metals* **2021**, *11*, 1957. https://doi.org/10.3390/met11121957

Received: 1 December 2021
Accepted: 2 December 2021
Published: 6 December 2021

Publisher's Note: MDPI stays neutral with regard to jurisdictional claims in published maps and institutional affiliations.

Copyright: © 2021 by the author. Licensee MDPI, Basel, Switzerland. This article is an open access article distributed under the terms and conditions of the Creative Commons Attribution (CC BY) license (https://creativecommons.org/licenses/by/4.0/).

Lesiuk et al. [6] showed a comparison of the results of the fatigue crack growth rate for raw rail steel, steel reinforced with composite material—CFRP—and the case of counteracting crack growth using the stop-hole technique, as well as with an "anti-crack growth fluid". It has been shown that the fatigue crack grows fastest in the case of the raw material and slowest in the case of the "anti-crack growth fluid" application. As a result of fluid activity, the fatigue crack closure occurred, which reduced the growth of this crack.

Ahmed et al. [7] investigated the fatigue crack propagation mechanism of CP Ti at various stress amplitudes. One crack at 175 MPa and three main cracks via sub-crack coalescence at 227 MPa were found to be responsible for the fatigue failure. The crack deflection and crack branching that cause roughness-induced crack closure (RICC) appeared at all studied stress amplitudes; hence, RICC at various stages of crack propagation (100, 300 and 500 μm) could be quantitatively calculated. Noticeably, a lower RICC was found at higher stress amplitudes (227 MPa) for fatigue cracks longer than 100 μm than for those at 175 MPa. This caused the variation in crack growth rates under the studied conditions.

Lee et al. [8] conducted fatigue tests at room temperature and 1000 K for 0.135-mm-thick alloy 625 tubes (outer diameter of 1.5 mm), which were brazed to the grip of the fatigue specimen. The variability in fatigue life was investigated by analyzing the locations of the fatigue failure, fracture surfaces and microstructures of the brazed joint and tube. At room temperature, the specimens failed near the brazed joint. Rusnak et al. [9] fatigue tested nine poles with 18 openings using four-point bending at various stress ranges. Among the 18 hand-holes tested, 17 failed in one way or another as a result of fatigue cracking. Typically, fatigue cracking would occur at either the three or nine o'clock positions around the hand-hole and then proceed to transversely propagate into the pole before failure. Finite element analysis was used to complement the experimental study.

Ishihara et al. [10] analyzed the structural integrity of ferritic steel structures subjected to large temperature variations, which required the collection of the fracture toughness (K_{Jc}) of ferritic steels in the ductile-to-brittle transition region. In this study, a Windows-ready K_{Jc} predictor based on tensile properties (specifically, yield stress and tensile strength at room temperature and yield stress at K_{Jc} prediction temperature) was developed by applying an artificial neural network to 531 K_{Jc} datapoints.

Liu et al. [11] subjected the 2524-T3 aluminum alloy to fatigue tests under the conditions of $R = 0$, a 3.5% NaCl corrosion solution and loading cycles of 10^6, and the S-N curve was obtained. The horizontal fatigue limit was 169 MPa, which is slightly higher than the longitudinal fatigue limit of 163 MPa. The influence mechanism of corrosion on the fatigue crack propagation of the 2524-T3 aluminum alloy was discussed. The fatigue source characterized by cleavage and fracture mainly comes from corrosion pits, whose expansion direction is perpendicular to the principal stress direction.

3. Conclusions and Outlook

In this Special Issue, there are various topics relating to the latest approach to fatigue crack growth. They relate to the influence of load, microstructure, friction, corrosion or to welded joints. However, many issues in this area of research have not yet been explored and the dissemination of these results should be continued. As a Guest Editor, I hope that the research results presented in this Special Issue will contribute to the further progression of research on the growth of fatigue cracks.

Finally, I would like to thank all the reviewers for their input and efforts in producing this Special Issue, and the authors for the papers they have prepared. I would also like to thank all the staff at the *Metals* Editorial Office, especially Toliver Guo, the Assistant Editor, who managed and facilitated the publication process.

Funding: This research received no external funding.

Conflicts of Interest: The author declares no conflict of interest.

References

1. Sharma, A.; Oh, M.C.; Ahn, B. Recent advances in very high cycle fatigue behavior of metals and alloys—A review. *Metals* **2020**, *10*, 1200. [CrossRef]
2. Wang, W.; Yang, J.; Chen, H.; Yang, Q. Capturing and micromechanical analysis of the crack-branching behavior in welded joints. *Metals* **2020**, *10*, 1308. [CrossRef]
3. Xu, W.; Zhao, Y.; Chen, X.; Zhong, B.; Yu, H.; He, Y.; Tao, C. An ultra-high frequency vibration-based fatigue test and its comparative study of a titanium alloy in the VHCF regime. *Metals* **2020**, *10*, 1415. [CrossRef]
4. Artola, G.; Aldazabal, J. Hydrogen assisted fracture of 30MnB5 high strength steel: A case study. *Metals* **2020**, *10*, 1613. [CrossRef]
5. Zakavi, B.; Kotousov, A.; Branco, R. The evaluation of front shapes of through-the-thickness fatigue cracks. *Metals* **2021**, *11*, 403. [CrossRef]
6. Lesiuk, G.; Nykyforchyn, H.; Zvirko, O.; Mech, R.; Babiarczuk, B.; Duda, S.; Farelo, J.M.A.; Correia, J.A.F.O. Analysis of the deceleration methods of fatigue crack growth rates under mode I loading type in pearlitic rail steel. *Metals* **2021**, *11*, 584. [CrossRef]
7. Ahmed, M.; Islam, M.S.; Yin, S.; Coull, R.; Rozumek, D. Fatigue crack growth behaviour and role of roughness-induced crack closure in CP Ti: Stress amplitude dependence. *Metals* **2021**, *11*, 1656. [CrossRef]
8. Lee, S.; Kim, H.; Park, S.; Choi, Y.S. Fatigue variability of alloy 625 thin-tube brazed specimens. *Metals* **2021**, *11*, 1162. [CrossRef]
9. Rusnak, C.R.; Menzemer, C.C. Fatigue behavior of nonreinforced hand-holes in aluminum light poles. *Metals* **2021**, *11*, 1222. [CrossRef]
10. Ishihara, K.; Kitagawa, H.; Takagishi, Y.; Meshii, T. Application of an artificial neural network to develop fracture toughness predictor of ferritic steels based on tensile test results. *Metals* **2021**, *11*, 1740. [CrossRef]
11. Liu, C.; Ma, L.; Zhang, Z.; Fu, Z.; Liu, L. Research on the corrosion fatigue property of 2524-T3 aluminum alloy. *Metals* **2021**, *11*, 1754. [CrossRef]

MDPI
St. Alban-Anlage 66
4052 Basel
Switzerland
Tel. +41 61 683 77 34
Fax +41 61 302 89 18
www.mdpi.com

Metals Editorial Office
E-mail: metals@mdpi.com
www.mdpi.com/journal/metals